Addressing Barriers to Low-Carbon Innovation

Finanzmärkte und Klimawandel

Herausgegeben von
Dirk Schiereck und Paschen von Flotow

Band 7

Friedemann Polzin

Addressing Barriers to Low-Carbon Innovation

Essays on Structures and Policies to Mobilise Private Finance

Bibliographic Information published by the Deutsche Nationalbibliothek
The Deutsche Nationalbibliothek lists this publication in the Deutsche
Nationalbibliografie; detailed bibliographic data is available in the internet
at http://dnb.d-nb.de.

Zugl.: EBS Universität für Wirtschaft und Recht, Diss., 2015

Library of Congress Cataloging-in-Publication Data
Polzin, Friedemann, 1985- author.
Addressing barriers to low-carbon innovation : essays on structures and policies to
mobilise private finance / Friedemann Polzin.
pages cm. — (Finanzmärkte und Klimawandel ; Band 7)
Includes bibliographical references.
ISBN 978-3-631-66981-5
1. Carbon dioxide mitigation—Economic aspects. 2. Green technology—Finance. 3.
Pollution prevention—Economic aspects. 4. Environmental policy—Economic aspects.
I. Title.
HC79.P55P65 2015
363.738'746—dc23

2015033304

D 1540
ISSN 2190-3069
ISBN 978-3-631-66981-5 (Print)
E-ISBN 978-3-653-06116-1 (E-Book)
DOI 10.3726/978-3-653-06116-1

Acknowledgements

This thesis benefited tremendously from the genuine support of Paschen von Flotow, executive director of the Sustainable Business Institute (SBI). Both his encouragement as well as his critique helped me designing, conducting and writing up this research. I would also like to thank Florian Täube, professor at Solvay School of Management and Economics, Brussels (formerly EBS Business School) who contributed significantly to the success of this thesis, especially during the academic publication process and by sharing his academic network. Moreover I would like to express my gratitude towards my external co-authors Laurens Klerkx, professor at Wageningen University and Colin Nolden, research fellow at SPRU (Science Policy Research Unit), University of Sussex. Finally I benefited from the ongoing constructive support of Ronald Gleich, head of the Strascheg Institute for Innovation and Entrepreneurship (SIIE) and professor at EBS Business School.

I gratefully acknowledge the generous funding I received from the German Federal Ministry for Education and Research (BMBF) through the research projects 'CFI – Climate Change, Financial Markets and Innovation' and 'Communication and Monitoring of Energy Service Contracts to accelerate the Diffusion of LED street lighting' as part of 'OLPOMETS'.

During the process of research many people spent their valuable time by sharing and discussing their perspective. I am very grateful for the time and support of many experts and decision makers during interviews and workshops. In addition, it was a great experience to discuss ongoing research at academic conferences and workshops which helped enormously to refine the focus and contribution of the journal articles. Especially the research visit in SPRU (University of Sussex, UK) was a fulfilling experience both in terms of research collaboration and personal development. I thank Prof Paul Nightingale, Prof Carlota Perez and Prof Marianna Mazzucato for sharing their knowledge, experience as researchers and 'big picture thinking'.

Furthermore I would like to thank my colleagues Christian, Conny, Daniel, Dennis, Marco and Ruhi at the SBI, Florian, Michael and all other colleagues at the SIIE as well as other EBS doctoral students for supporting me and making me feel at home. The same applies to my temporary colleagues in SPRU Edwin, Tomas and Veronica. I would also like to acknowledge many of my friends in Germany who supported this thesis by helping me in focussing on my research or by occasionally advising me to relax.

Finally I am sincerely indebted to my family, both my parents, Gerlinde and Andreas as well as my brother Georg, who encouraged me to pursue my own ideas and get involved into the political and scientific debate. Their emotional support and understanding have helped me very much to complete this thesis.

Table of contents

List of figures

List of tables

List of abbreviations

ARPA-E	Advanced Research Projects Agency – Energy
ATP	Advanced Technology Program
BA	Business angel
BNEF	Bloomberg New Energy Finance
EPC	Energy performance contracting
ESCo	Energy service company
ETS	Emission trading system
EU	European Union
FIT	Feed-in tariff
GHG	Greenhouse gas
IEA	International Energy Agency
IPCC	Intergovernmental Panel on Climate Change
LED	Light emitting diode
MLP	Multi-level perspective
NGO	Non-governmental organisation
OECD	Organisation for Economic Co-operation and Development
PE	Private equity
PPP	Public private partnership
R&D	Research and development
RPS	Renewable portfolio standard
RD&D	Research, development and demonstration
RE	Renewable energy
ROC	Renewable obligation certificates
RQ	Research question
SBIR	Small Business Innovation Research
STI	Science, technology and innovation
TC	Transaction cost
TCE	Transaction cost economics
VC	Venture capital

1 Introduction

1.1 Motivation, purpose and scope of the research

Global climate change has been recognised amongst the biggest 'grand challenges' facing humanity in the 21st century. It has been agreed that this relates to growing CO_2 emissions, caused by human activity linked to further megatrends, such as growing populations and sectoral developments. There is widespread consensus among policy makers, businesses, the scientific community and wider society that the transition towards a low-carbon economy (i.e. 'green economy') is the desirable goal, which makes a sustainable life possible for all human beings by decoupling economic activity from the use of finite resources (Flotow & Schiereck, 2013; IEA, 2013b; IPCC, 2014; Marcucci & Turton, 2013; OECD, 2009; Popp, 2012; Rosen & Guenther, 2014; Stern, 2007, 2008).

Climate change impact is becoming more severe across the world, with developing countries currently being affected the most. However, Europe also witnesses a relevant impact today, one that is projected to increase in the medium- to long-term future. Facing these challenges, Europe has established ambitious targets regarding the reduction of CO_2 emissions (i.e. 20-20-20 targets[1] and an 80% reduction of CO_2 emissions compared to 1990 by 2050). On the one hand, these circumstances call for a strategy to achieve lower carbon emissions effectively and efficiently. On the other, governments across the world are currently preoccupied with addressing the more tangible financial crisis and its accompanying short-term problems. The financial crisis has resulted in risk aversion among private financiers, thus hindering investments in the real economy, especially clean technologies. In addition it has also aggravated the budgetary crises in Europe. Both aspects threaten green growth in the future. Consequently, climate change and the financial crisis are economically interrelated.

To target both crises at once, scholars suggest a 'green growth' approach, i.e. achieving sustainable development and remaining competitive in a globalised world (Flotow & Schiereck, 2013; Mazzucato & Perez, 2014). Policy makers have already translated this vision into the 'smart, sustainable and inclusive growth strategy', supported by the EU Eco-innovation plan (European Commission,

---

1   The '20-20-20' targets set three key objectives for 2020: A 20% reduction in EU greenhouse gas emissions from 1990 levels; An increase in the share of EU energy consumption produced from renewable resources to 20%; And a 20% improvement in the EU's energy efficiency (European Commission, 2009).

2010). This thesis aims at providing evidence based on cases that tackle both these crises on a tangible, technology specific level which then yield the opportunity for green growth and a transition towards a 'green economy'.

A critical element to achieve green growth is the development and diffusion of clean technologies (Foxon, Köhler, & Oughton, 2008; Hargadon, 2010; IPCC, 2014; Mowery, Nelson, & Martin, 2010). This process is hampered by a number of factors, relating both to the inherent characteristics of innovation and technological change, and environmental externalities (Foxon & Pearson, 2008; Jaffe, Newell, & Stavins, 2005). Scholars investigated possible market-based, regulatory and innovation policy instruments to address these failures on an abstract level, as well as with relation to concrete technologies and contexts, to derive policy implications (M. A. Brown, 2001; Gallagher, Holdren, & Sagar, 2006; Leete, Xu, & Wheeler, 2013; Leitner, Wehrmeyer, & France, 2010; Mowery et al., 2010).

Among the most salient barriers to the commercialisation and diffusion of clean technologies, scholars have highlighted the financing environment (Iyer et al., 2013; Jacobsson & Karltorp, 2013; Leete et al., 2013; Zhang, Shen, & Chan, 2012). Ideally, the financial market allocates resources to ensure that economic growth is possible. However, this does not necessarily lead to the social optimum with regard to clean technologies. On the one hand, recent investment trends show decreasing amounts of finance dedicated to clean technologies and correspondingly, an increasing risk aversion of financiers (BNEF, 2013; Flotow & Schiereck, 2013). On the other, clean technologies require huge amounts of investments in companies, projects and infrastructure (Mathews, Kidney, Mallon, & Hughes, 2010; Perez, 2013). Yet there is a surprisingly little amount of research in this area beyond the classical innovation finance stream of venture capital (VC) (Kenney & Hargadon, 2012; Marcus, Malen, & Ellis, 2013; Olmos, Ruester, & Liong, 2012). This thesis therefore focuses on, and contributes to, a more holistic understanding of the peculiarities that eco-innovations face with regard to finance. This perspective proves beneficial both for policy makers and for theorists (Bürer & Wüstenhagen, 2009; Jefferson, 2008; Mathews et al., 2010a; Perez, 2013; van den Bergh, 2013).

The context of this thesis centres on questions of how finance flows can be guided towards innovative and sustainable projects, infrastructure and companies that create low-carbon value and growth for future generations. The individual chapters analyse structures (i.e. identifiable organisational, institutional or legal arrangement between actors in the innovation system) and policy environments that are conducive to private investments. Early and late stage innovations require continuous flows of finance to overcome structural gaps in the innovation process, such as the 'valley of death' and transitions to achieve green growth.

1.2 Theoretical background and research context

Eco-innovation refers to the development, commercialisation and diffusion of innovative or novel technologies that reduce carbon emissions and other negative environmental effects (Foxon et al., 2008; Iyer et al., 2013)[2]. Low-carbon innovation exhibits peculiarities in comparison with conventional innovation, such as the double externality problem due to incomplete markets for technologies and emissions (Jaffe et al., 2005; Rennings, 2000).

Thus eco-innovations suffer from market and system failures resulting from interrelated technological, institutional and economic lock-ins and path dependencies in fossil-fuel based technologies (Coenen & Diaz Lopez, 2010; Foxon et al., 2005; Foxon & Pearson, 2008). Examples include the inferior performance of clean technologies and competence problems of innovators and users (Bergek & Onufrey, 2013; Chadha, 2011; Foxon et al., 2008; Iyer et al., 2013), information asymmetries (imperfect information) and bounded rationality (Foxon & Pearson, 2008; Jaffe et al., 2005; Klein Woolthuis, Lankhuizen, & Gilsing, 2005; Sanstad & Howarth, 1994) as well as the externalities and investments into fossil-fuel based innovations in the past (Foxon & Pearson, 2008; Jaffe et al., 2005; Jaffe & Stavins, 1994; Rennings, 2000; Wüstenhagen & Menichetti, 2012). In addition, niche-regime dynamics that favour fossil-fuel-based technologies prevent the transition towards sustainable innovation (Hekkert, Suurs, Negro, Kuhlmann, & Smits, 2007; Nill & Kemp, 2009; A. Smith, Voß, & Grin, 2010).

In order to account for these deficits, policy makers developed a variety of responses to heal market and systemic failures. Generally, the combination of subsidies with taxes and regulation, technology push and demand pull policies (Kemp & Oltra, 2011; Loiter & Norberg-Bohm, 1999; Veugelers, 2012) as well as internalization of externalities through greenhouse gas (GHG) emission trading or GHG taxation is suggested (Fischer & Newell, 2008; Popp, 2010; Rogge, Schneider, & Hoffmann, 2011). Maintaining a focus on competitiveness and a sustainable transition of society by connecting market formation and policy incentives is crucial (Coenen & Diaz Lopez, 2010; Dewald & Truffer, 2011; Szabó & Jäger-Waldau, 2008). Hence, efficient policy making requires information about technologies, external effects and economic interdependencies.

Combinations of different policies tailored to different stages in the innovation cycle and different markets assure an effective and efficient dissemination of novel

---

2   Eco-innovation, low-carbon innovation, environmental innovation or innovation in clean technologies are used interchangeably throughout this thesis. See also chapter 2 for a detailed definition.

technologies (Foxon et al., 2008; Hargadon, 2010; Mowery et al., 2010). Thereby, the dissemination can be divided into three distinct phases: Technology generation (Phase I), technology commercialisation (Phase II), and technology diffusion (Phase III). Phase I can further be divided into basic and applied research and development (R&D), Phase II into demonstration and pre-commercial and Phase III into niche-market, supported commercial and fully commercial (Auerswald & Branscomb, 2003; Bürer & Wüstenhagen, 2009; Wüstenhagen & Menichetti, 2012) (Figure 1-1).

Figure 1-1: Stylised innovation cycle for clean technologies

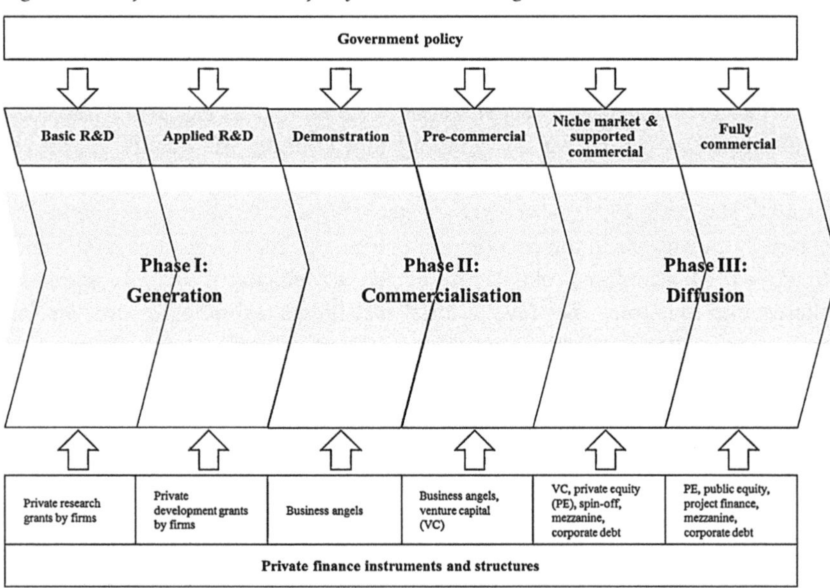

Adapted from Auerswald & Branscomb (2003); Bürer & Wüstenhagen (2009); Wüstenhagen & Menichetti (2012).

The stylised development and diffusion processes of clean technologies represent complex interdependent phenomena which involve public and private actors. Both take different roles along the various stages (Auerswald & Branscomb, 2003; Wüstenhagen & Menichetti, 2012). During the basic and applied R&D stages (Phase I) technologies are being developed by both public (research institutes, universities) and private organisations (firms) which supply the necessary financial resources in the form of public or private research grants and subsidies. Due to limited appropriability and other externalities, that result in product and market uncertainty, private investments in R&D remain below the social optimal level (i.e. under-investment) (Barreto & Kemp, 2008; Montalvo, 2008; Mowery et al., 2010; Nemet & Kammen, 2007). Thus typically, public R&D subsidies and tax-credits far outweigh the private sector engagement (Olmos et al., 2012).

At the end of the applied R&D stage and during the commercialisation phase (Phase II), governments typically deploy a range of technology push instruments, such as funding public private research partnerships and demonstration or pilot projects, and demand pull measures, for example by providing the necessary infrastructure and regulations to overcome this 'valley of death' (Chadha, 2011; Hendry, Harborne, & Brown, 2010; Jacobsson & Bergek, 2004; Jacobsson & Lauber, 2006; Kenney & Hargadon, 2012). At this point, the private sector begins a (limited) engagement. Business angels (BA) and venture capitalists (VC) as private financiers invest into start-ups and small innovative firms, whereas large or mature firms start to deploy internal funds for ongoing R&D and commercialisation activities. During the pre-commercial stage, the technology is usually sufficiently mature to allow for scale up towards production which is financed by VCs (Kenney & Hargadon, 2012; Marcus et al., 2013).

Beginning with the niche-market stage (Phase III), ideally the private sector actors take the lead to foster diffusion of the technology. Firms concentrate on market development. Banks, private equity investors and internal funds provide the necessary resources to finance production and marketing (B. H. Hall, 2002; B. H. Hall & Lerner, 2010). Governments adjust the institutional environment to minimize regulatory and political risks (M. A. Brown, 2001; Chassot, Hampl, & Wüstenhagen, 2014; Haley & Schuler, 2011; Oltra & Jean, 2005).

Throughout the different phases clean technologies face particular technological, institutional, and economic barriers that exhibit financial aspects. These result from the interplay between private actors and governmental engagement in the form of science, technology and innovation (STI) policy and regulation along the innovation process (Foxon et al., 2008; Mathews et al., 2010; Wüstenhagen & Menichetti, 2012). However, to achieve the transition towards a low-carbon society, significant levels of investment are needed to make new technologies competitive with incumbent technologies (Foxon et al., 2008). Financial capital plays a crucial role as it supports innovation (Perez, 2004).

In the case of eco-innovation, the relationship between finance and innovation is more complex. In fact, clean technologies exhibit higher uncertainty, regulatory dependency and capital intensity, which makes them unattractive for private financiers as these possess limited abilities to screen potential targets (Kenney & Hargadon, 2012; Randjelovic, O'Rourke, & Orsato, 2003). In consequence, this leads to a thin financial market especially for eco-innovation all along the innovation cycle (Dahlstrand & Cetindamar, 2000; Nightingale et al., 2009). Accordingly, the investments into R&D, commercialisation and diffusion of clean technologies remain below the socially desirable level (Hargadon, 2010; Mathews et al., 2010; Mowery et al., 2010).

1.3 Research questions and methodologies

An overview of the structure and logic of arguments within this thesis can be drawn from Figure 1-2. Both pillars, i.e. structures and policies, prove practically and theoretically relevant to mobilise private finance along the innovation cycle of low-carbon technologies. The selected cases offer evidence of how to analyse the structural and policy deficit with regard to financing low-carbon innovation. They provide insights from a variety of technologies that are needed to transition towards a 'green economy', notably energy efficiency technologies, renewable energy technologies and other clean technologies such as bio-economy, e-mobility and smart grids.

Figure 1-2: Structure of this thesis

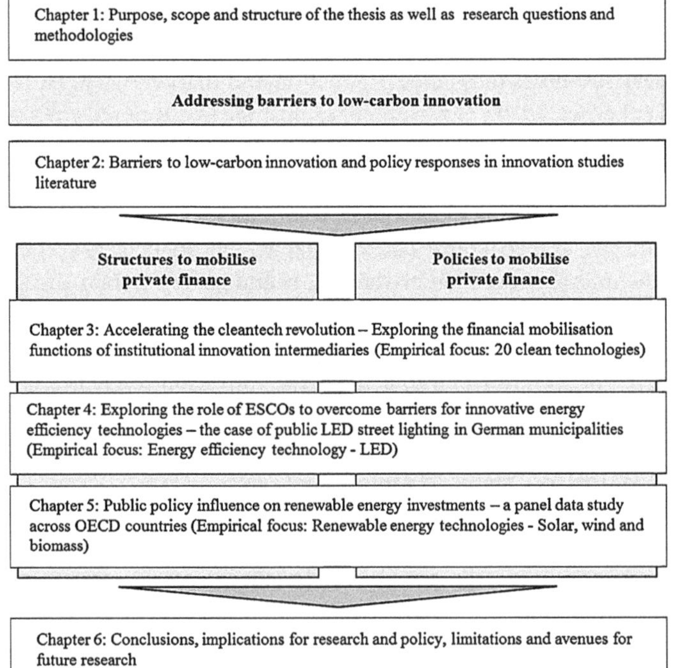

Chapter 2 reviews the relevant body of innovation studies literature and contributes to the discussion by integrating previously separated literature streams of innovation systems (IS), environmental economics, energy economics, energy policy and transition studies to identify barriers to low-carbon innovation as well as possible policy measures to overcome them (Fischer & Newell, 2008; Jacobsson & Bergek, 2011; Jakeman, Hanslow, Hinchy, Fisher, & Woffenden, 2004; Markard, Raven, & Truffer, 2012; Newell, Jaffe, & Stavins, 2006; Popp, 2010). The chapter organises the barriers and policy measures according to the technology innovation cycle for clean technologies (see Figure 1-1).

Failures along the innovation cycle stem from technological, institutional, economic, financial, political, interaction, capability and transformational barriers. Accordingly, policy makers possess a range of options through active technology policy, institutional support, market creation, mobilising finance, policy design, fostering interaction, increasing knowledge and learning as well as facilitating transformation (Bürer & Wüstenhagen, 2009; Chadha, 2011; Hargadon, 2010; Kenney & Hargadon, 2012; Leitner et al., 2010; A. Smith et al., 2010; van den Bergh, 2013; Weber & Rohracher, 2012). A combination of financial, economic, institutional and transitional barriers slows down the development, commercialisation and diffusion of clean technologies (Iyer et al., 2013). This in turn inhibits politically-induced transition processes. Therefore the subjacent research questions (RQ) read as follow:

RQ1a: *How do abstract failures become manifest in specific barriers to eco-innovation along the innovation cycle?*
RQ1b: *What possibilities exist for policy makers to overcome them?*

Chapter 3 analyses examples of specific structures to mobilise private investments at critical stages in the innovation cycle. It focuses on the transition between applied R&D phase, demonstration and pre-commercial phases. Barriers early in the innovation cycle revolve around knowledge spillovers, the missing capabilities and networks of firms as well as missing university-industry links (Gallagher et al., 2006; Iyer et al., 2013). These are typically addressed by establishing strategic research partnerships (Audretsch, Link, & Scott, 2002; Kenney & Hargadon, 2012; Sovacool, 2008). However, major obstacles to eco-innovation severely affect the commercialisation phase which directly follows. These include infrastructure problems, and finance-related problems, such as high capital intensity and limitations of early stage private financing (Jacobsson & Karltorp, 2013; Köhler, Wietschel, Whitmarsh, Keles, & Schade, 2010; Leete et al., 2013; Marcus et al., 2013).

To address this phase in the innovation cycle and to account for the diversity of possible solutions, a qualitative research design was applied. Following the recommended approach to select case studies for analytic induction, theoretical sampling was used (Eisenhardt, 1989; Siggelkow, 2007; Yin, 2009). The chapter focuses on 20 cleantech R&D partnerships in Germany from all relevant project managing organisations in the framework of 'Research for Sustainable Development'. It builds on interviews with 25 R&D project managers that are embedded into an archival document analysis. The partnerships cover a variety of technologies such as energy efficiency, energy production, storage, advanced materials and biotechnology from different sectors including chemicals, energy, agriculture and mobility. Thereby the chapter reveals the importance of innovation intermediaries between public and private actors in strategic research partnerships as possible agents to partially address the aforementioned system failures (Howells, 2006; Kivimaa, 2014; Klerkx & Leeuwis, 2009). However, the functions of intermediaries to address finance-related barriers and to leverage public R&D investments have been under-researched. Hence the research question focuses on the following:

RQ2: How do institutional innovation intermediaries address the complex set of barriers surrounding (low-carbon) innovation from R&D to commercialisation?

Chapter 4 addresses the critical demonstration, pre-commercial and niche-market or supported commercial phases. In addition to most barriers of the demonstration phase prevailing, these barriers revolve around infrastructure, technological and institutional lock-in, and other regulatory risks, as well as market and demand articulation (Bergek, Mignon, & Sundberg, 2013; Foxon et al., 2008; Haley & Schuler, 2011; Iyer et al., 2013; Köhler et al., 2010; Wieczorek & Hekkert, 2012). Drawing from the literature on science, technology and innovation (STI) policy, this chapter explores the effects of public procurement on the commercialisation of LED street lighting, by mobilising private investments on the municipal level (Edler & Georghiou, 2007; Edquist & Zabala-Iturriagagoitia, 2012; Hartmann, Roehrich, Frederiksen, & Davies, 2014; Uyarra, Edler, Garcia-Estevez, Georghiou, & Yeow, 2014). Focusing on LED street lighting as an innovative energy efficiency technology, the chapter argues that cooperation between public municipalities and private energy service companies (ESCos) could address barriers. Energy performance contracting (EPC) represents a public-private structure and finance mechanism to operationalise this cooperation (Hannon, Foxon, & Gale, 2013; Sorrell, 2007).

To understand the relationship between the innovation process of an energy efficiency technology and energy service contracts, the intersection of LED and ESCo markets is analysed as a case study (Miles & Huberman, 1999; Patton, 2002).

The chapter explores this intersection by focusing on the application of LED street lighting in German municipalities in a longitudinal, multi-level analysis of key market players and key policy initiatives as a research context. A research design that integrates an extensive longitudinal archival document analysis and interviews permits reflection upon changes in the industry and institutional context (Seawright & Gerring, 2008). Accordingly the research question is as follows:

RQ3: What is the role of ESCO solutions in addressing the barriers to commercialisation and diffusion of LED street lighting in Germany?

Chapter 5 then investigates the direct effect of policy instruments on investments in renewable energy (RE) technologies across a set of OECD countries in a longitudinal panel analysis. This chapter addresses the later stages in the innovation cycle for clean technologies (i.e. niche-market, supported commercial and fully commercial). It focuses on economic and institutional lock-ins as well as regulatory uncertainty in RE production (Bergek et al., 2013; Jacobsson & Karltorp, 2013; Leete et al., 2013; Zhang et al., 2012). By estimating the most effective policy measures to encourage RE investments by institutional investors such as banks, private equity and pension funds, the chapter focuses on project and asset finance structure to diffuse mature RE technologies (Bergek et al., 2013; Wüstenhagen & Menichetti, 2012).

To analyse the effectiveness of different policy instruments, a quantitative research design was applied. Investigating the diffusion of a particular technology and corresponding investments requires a longitudinal research design (Aguirre & Ibikunle, 2014; Angrist & Pischke, 2008; Hair, 2010; Johnstone, Haščič, & Popp, 2010). A large number of countries were considered to examine patterns of policy measures followed by investments. To provide generalizable results, the chapter covers a variety of OECD countries, thus conducting a panel data regression (Marques & Fuinhas, 2012a; Wooldridge, Calhoun, Jung, Greber, & Montgomery, 2009). The corresponding research question reads as follows:

RQ4: Which policies have proven (most) conducive to investments in renewable energy assets?

The remainder of this thesis is structured as follows: Chapter 2 reviews the diverse body of literature dealing with barriers to low-carbon innovation and policy responses and identifies gaps in the literature. Chapter 3 and Chapter 4 focus on structures in the context of public-private-cooperation that mobilise private finance at the commercialisation and diffusion stages. Chapter 3 to 5 additionally investigate policies that are conducive to attracting private finance in these stages. Chapter 6 summarises the findings and draws implications for policy makers and researchers, outlines some limitations of the research and possible future work. Table 1 provides an overview about the chapters of this thesis with corresponding research questions, methods, cases and data sources.

Table 1-1: Overview about journal articles as chapters in this thesis

Chapter / Journal article	Research question(s)	Research method / Case	Data source
Chapter 2: *Barriers to low-carbon innovation and policy responses in innovation studies literature*	How do abstract failures become manifest in specific barriers to eco-innovation along the innovation cycle? (RQ1a) What possibilities exist for policy makers to overcome them? (RQ1b)	Narrative literature review	Literature on Eco-/low-carbon/ cleantech innovation (1994–2014)
Chapter 3: *Accelerating the cleantech revolution – Exploring the financial mobilisation functions of institutional innovation intermediaries*	How do institutional innovation intermediaries address the complex set of barriers surrounding (low-carbon) innovation from R&D to commercialisation? (RQ2)	Observational/ quali-tative case study (Cross-sectional) 20 government-supported R&D cleantech partnerships in Germany	Archival documents 25 interviews with R&D project managers (2012) 1 workshop with 20 study participants (2013)
Chapter 4: *Exploring the role of ESCOs to overcome barriers for innovative energy efficiency technologies – the case of public LED street lighting in German municipalities*	What is the role of ESCO solutions in addressing the barriers to commercialisation and diffusion of LED street lighting in Germany? (RQ3)	Observational/ qualitative case study (Longitudinal) LED and ESCo markets in Germany	Archival documents (2008–2013) 40 interviews with public and private actors along the value chain (2014)
Chapter 5: *Public policy influence on renewable energy investments – a panel data study across OECD countries*	Which policies have proven most conducive to investments in renewable energy assets? (RQ4)	Quantitative / panel data analysis (Longitudinal) Solar, Wind, Biomass sectors across OECD countries	Investment data – Bloomberg New Energy Finance (2003–2012) Policy support and measures data – IEA (2000–2012)

2 Barriers to low-carbon innovation and policy responses in innovation studies literature

Friedemann Polzin

Abstract: This paper analyses the field of innovation studies i.e. innovation systems, transition literature, environmental and energy economics regarding barriers for the commercialisation and diffusion of clean technologies (low-carbon innovation). The barriers are organised along the innovation cycle, through which this article attempts to integrate previously separated literatures and bridge the gap between abstract failures and tangible barriers to allow for more differentiated policy responses. Results show that failures along the innovation cycle stem from technological, institutional, economic, financial, political, interactional, capability and transformational barriers. Policy makers aiming to tailor their responses accordingly possess a range of options through active technology policy, institutional support, market creation, mobilisation of financial resources as well as fostering interaction, increasing knowledge and learning, facilitating transformation and overall policy design criteria. One final conclusion revolves around future research focused on financing eco-innovation and corresponding policy mechanisms such as public-private cooperation.

2.1 Introduction

Climate change has been acknowledged as one of the biggest challenges to mankind by the wider society and a range of academic disciplines, including natural sciences, economics and political science (IPCC, 2014; Rosen & Guenther, 2014; Stern, 2008). Scholars have come up with ways to reduce the emissions that cause climate change, one of which is the development, commercialisation and diffusion of cleaner technologies (Foxon & Pearson, 2008; Iyer et al., 2013; van den Bergh, 2013). However, systemic failures do exist, for example the environmental public good problem in the absence of carbon markets and failing markets for technologies. Additionally, the time-critical transformation of relevant markets translates into huge costs and public-private uncertainties (Godoe & Nygaard, 2006; Huberty & Zysman, 2010; Mowery et al., 2010). Thus addressing the barriers and accelerating the innovation process while maintaining a functioning market is crucial.

Eco-innovation has been researched from a variety of perspectives including innovation systems (IS) (Jacobsson & Bergek, 2011), transition studies (Markard et al., 2012), environmental and ecological economics (Böhringer, Mennel, &

Rutherford, 2009; Fischer & Newell, 2008; Newell et al., 2006), as well as energy economics and policy (Jakeman et al., 2004; Popp, 2010). To gain a holistic picture of the technological, economic and institutional processes surrounding eco-innovation, an interdisciplinary approach is adopted thereby enabling the integration of literature and debate streams, which have previously been separated. These streams and main constructs, reconciled under the umbrella of innovation studies[3], are depicted in Figure 2-1.

Figure 2-1: Literature streams and main concepts

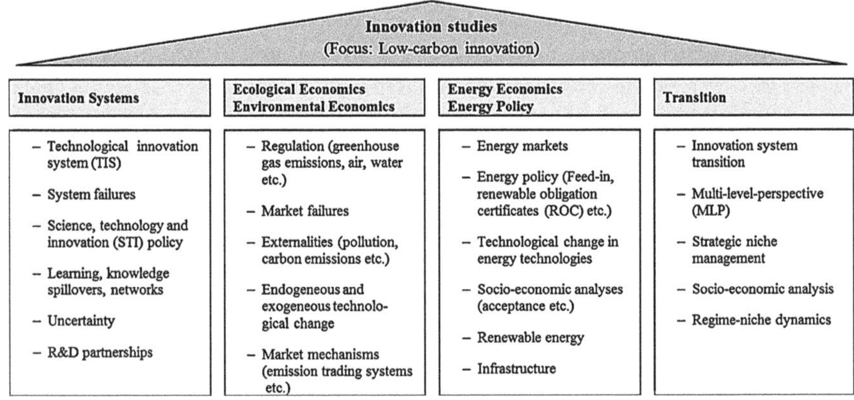

Newell et al. (2006, p. 563) state that 'a particularly important and understudied aspect [...] is the conceptual and empirical modelling of how the various stages of technological change are interrelated, how they unfold over time, and the differential impact that various policies [...] may have on each phase of technological change.' Following this argument, which has been put forward by other scholars (e.g. van den Bergh, 2013) the objectives of this literature review are threefold:

---

3   These include economics of innovation with a focus on environmental and energy economics, innovation systems (including science, technology and innovation (STI) policy), innovation management and transition (Martin, 2012b; Martin, Nightingale, & Yegros-Yegros, 2012).

1. *To review high-quality published literature revolving around systemic or market failures for low-carbon innovation and organise it along the innovation process,*
2. *to identify specific barriers that relate to (abstract) market failures as well as policy responses and*
3. *to identify empirically under-researched areas.*

The review is structured as follows: Chapter 2.2 describes the methodology used to assemble the literature base, which is then reviewed in the following chapters. Chapter 2.3 draws on a process framework for eco-innovation and organises barriers, obstacles and challenges for eco-innovation accordingly. Chapter 2.4 addresses the preceding and possible policy responses. Finally, chapter 2.5 discusses implications for future research and policy.

2.2 Methodology

2.2.1 Research approach

The methodology is based on a traditional narrative review that seeks to understand a particular body of literature (Hart, 1998). More sophisticated methods exist that quantify evidence, making it possible for example to infer from the data in a meta-analysis (e.g. David & Han, 2004; Hunter & Schmidt, 2004; Pittaway, Robertson, Munir, Denyer, & Neely, 2004) or to make connections between clusters of authors in a bibliometric analysis (e.g. Babl, Schiereck, & Flotow, 2012). However, the methodological approach has deliberately been kept simple, in order to portray a relatively broad topic and do justice to the expected heterogeneity in the literature. Essentially the goal was to identify a representative base of articles that describe the barriers to low-carbon innovation and possible policy solutions. It is neither intended to be comprehensive, nor does it ignore critical theoretical perspectives. The articles were identified and classified, the texts analysed and finally mapped into a theoretical framework that provided a background for the research synthesis (Hart, 1998).

To understand the structure and as a starting point to address this heterogeneous body of literature, systemic review articles in the field were identified. These articles provide a 'map of the field' of innovation studies (Fagerberg, Fosaas, & Sapprasert, 2012; Martin, 2012b; Martin, Nightingale, & Yegros-Yegros, 2012). They were synthesised, the main barriers to eco-innovation outlined from the body of literature that surrounds these underlying barriers. The literature search applied certain criteria:

The first choice was to include only published seminal books, established and well-regarded working paper series (i.e. NBER, CEPR) and peer-reviewed articles that have been ranked A, B, C or D by the Association of University Professors of Business in German speaking countries (Verband der Hochschullehrer für Betriebswirtschaft – VHB)[4]. Doing so assured the compiled research achieved a certain level of quality. According to Hunter & Schmidt (2004) this does not lead to an 'availability bias' for empirical studies because if the number of articles is sufficiently large, the direction of the results published and those not published tend to be the same. The second choice was to use five scientific search engines that are widely used in the community of business scholars to carry out keyword searches. The search engines reviewed include Business Source Complete, Science Direct, EBSCO, Emerald and Google Scholar.

After identifying the main articles dealing with the barriers and imperfections related to eco- innovation, the articles have been analysed in a narrative review, revealing key concepts and central contributions (i.e. the 'knowledge base') behind the rationales for these failures (Fagerberg et al., 2012). The analysis followed these steps:

1. The analysed barriers regarding innovation have been synthesised and listed, as well as possible policy solutions if provided in the articles have grouped accordingly
2. Categories of barriers and policy responses were developed from the system failures literature on IS in the context of sustainability (Edquist, 2011; Jacobsson & Bergek, 2011; Klein Woolthuis et al., 2005; Weber & Rohracher, 2012)
3. These categories formed the basis for a theoretical framework to address challenges in the context of eco-innovation

A database containing authors, title, publication, main argument, chain of arguments, empirical or conceptual setting as well as keywords was developed. According to the main argument and keywords section, the articles have been classified for better organisation (Hart, 1998). Sections were written as the articles relevant to particular themes were reviewed.

---

4   A ranking developed on behalf of the Association of University Professors of Business in German speaking countries (Verband der Hochschullehrer für Betriebswirtschaft – VHB-JOURQUAL 2.1). The only exception consists of articles published in the journal of Environmental Innovations and Societal Transitions which is a new high-quality journal in the field of low-carbon innovation, not yet ranked in the VHB Jourqual 2.1.

2.2.2 Developing an analytical framework for the literature review

Based on previous literature (Foxon & Pearson, 2008; Horbach, Rammer, & Rennings, 2012; Rennings, 2000) this article adopts the following definition for the main concept: Low-carbon innovation[5] can be defined as the 'invention, commercialisation and diffusion of technologies that reduce carbon emissions and/or other environmentally negative impacts and thus contributes to sustainability'. Eco-innovation takes the form of product, process, service, organisational (e.g. business model), institutional or systemic innovation, the last being a combination of the aforementioned. It also exhibits characteristics of disruptive innovation (Mowery et al., 2010), and continuity-led or incremental innovation (Hargadon, 2010).

The phenomenon of low-carbon innovation has been studied through a variety of lenses. Drawing from the environmental economics literature, which emerged in accordance with the creation of specific journals throughout the 1990s, scholars began to focus on the role of technological change and technology diffusion for environmentally friendly technologies based on neoclassical market failure arguments (Jaffe & Stavins, 1994, 1995). These approaches have fallen short in analysing systemic problems and interdependencies in the specific national or technological innovation systems. Rennings (2000) highlights three peculiarities of eco-innovation. First, these specific innovations are subject to the double externality effect (knowledge spillovers and environmental externalities). Economically, the social return, e.g. climate change mitigation and knowledge spillovers, far exceeds the private return which is uncertain due to missing international climate conventions (Popp, 2010). Second, the regulatory push/pull effect to induce innovation is especially strong. Third, the technological innovations exhibit an increasing importance of accompanying social and institutional innovation to overcome institutional lock-in associated with changing patterns of behaviour social rules and norms. Rennings (2000) further highlights the importance of evolutionary approaches to consider the complexity of factors surrounding the failures. This IS stream of literature which followed his argument identified either abstract frameworks for policy makers or explored concrete instruments while taking a systemic perspective (Foxon et al., 2008; Foxon & Pearson, 2008; Hekkert & Negro, 2009; Jacobsson & Bergek, 2011). The aim of this literature review is to connect the abstract market and system failures literature (i.e. economics and IS)

---

5   Throughout the course of the analysis eco-innovation, low-carbon innovation, innovation in clean technologies and environmental innovation will be used interchangeably.

with concrete context dependent literature on barriers to eco-innovation that target specific phases of the innovation cycle.

To achieve this aim, firstly this article uses a stylised innovation cycle framework to organise the heterogeneous barriers according to the stage in which they occur, hence establishing a temporal perspective. This framework has been used to address finance gaps and institutional problems in the early stages of the innovation process in general and forms a basis to develop a conceptual perspective on the process for low-carbon innovation (Auerswald & Branscomb, 2003; Bürer & Wüstenhagen, 2009; Wüstenhagen & Menichetti, 2012). It allows for a differentiated view and corresponding policy actions.

Secondly, this article employs the IS framework consisting of actor networks and institutions on a technological level (TIS) (Bergek, Jacobsson, Carlsson, Lindmark, & Rickne, 2008; Carlsson & Stankiewicz, 1991; Hekkert et al., 2007; Jacobsson & Bergek, 2004; Malerba, 2002). This perspective has been found suitable for an innovation-based analysis of transitions towards a low-carbon economy based on clean technologies (Hekkert & Negro, 2009; Jacobsson & Bergek, 2011; Markard et al., 2012; A. Smith et al., 2010). There is consensus in the literature that a TIS serves the invention, commercialisation and diffusion of particular technologies.

However, this function is hampered by a number of problems that are described in the system failures framework (Bleda & del Río, 2013; Jacobsson & Bergek, 2011; Klein Woolthuis et al., 2005; Weber & Rohracher, 2012). These include technological, institutional (e.g. infrastructure, lock-in effects), economic (e.g. externalities, resource mobilisation), financial, political (e.g. coordination, reflexivity), interaction (e.g. networking, cooperation) and capability failures (e.g. knowledge, learning). Building on system failures and an extension of the IS theory with a stronger process focus, Jacobson and Bergek (2011) and Hekkert and Negro (2009) have demonstrated the ability of the TIS framework to identify a diverse set of system weaknesses in the area of environmental innovations. It allows policy makers to identify the points in a system where intervention is likely to matter the most. Consequently this failure categorisation serves as a basis throughout the review.

2.3 Barriers to low-carbon innovation along the innovation cycle

To incorporate the different literature streams selected for analysis, a broad definition for a barrier to low-carbon innovation was applied. A 'barrier' is defined as a blocking mechanism, obstacle or hampering mechanism, that prevents clean technologies from being commercialised and diffused which in turn inhibits technological change (Foxon & Pearson, 2008; Jacobsson & Karltorp, 2013). An overview about the categories of barriers organised along the innovation cycle which are discussed below can be drawn from Figure 2-2. The resulting framework is a combination of earlier work and comes with limitations regarding boundaries between the single barriers and potential overlaps (Bleda & del Río, 2013; Bürer & Wüstenhagen, 2009; Klein Woolthuis et al., 2005; Wüstenhagen & Menichetti, 2012).

Figure 2-2: Barriers to commercialisation and diffusion of eco-innovation

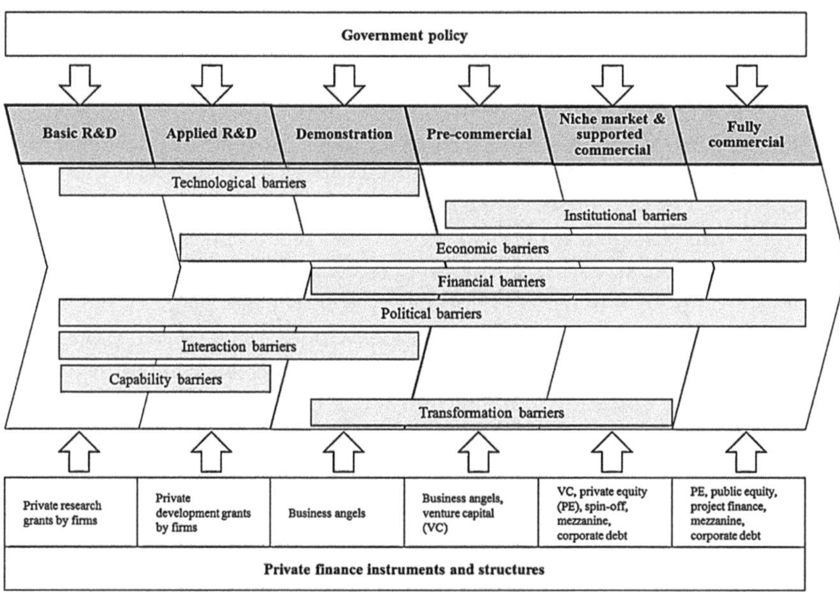

Own representation. Framework adapted from Auerswald & Branscomb (2003); Bürer & Wüstenhagen (2009); Wüstenhagen & Menichetti (2012).

2.3.1 Technological barriers

The overarching technological barrier facing innovative clean technologies is technological lock-in and path dependency (Bergek & Onufrey, 2013; Foxon & Pearson, 2008; Iyer et al., 2013; Schmidt & Marschinski, 2009) which relates to insufficient technological maturity, missing standards or infrastructure (e.g. electric vehicles compared to conventional vehicles). These developments are persistent due to suboptimal investments in clean R&D which leads to path dependency.

In the early technology generation stages (basic and applied R&D), one can attribute these abstract failures to knowledge deficits and learning below the optimal level as well as low spatial diffusion of knowledge (Barreto & Kemp, 2008). Beginning with the demonstration phase, stakeholders start to perceive technological risks and complexity associated with these new clean technologies such as long-term performance and effects on the socio-economic and natural environment (Böhringer et al., 2009; Iyer et al., 2013; Masini & Menichetti, 2012; Schleich, 2009; Wüstenhagen & Menichetti, 2012). Similarly reverse salients i.e. unanticipated political, economic and social consequences could hinder further development and commercialisation of a particular technology (Gee & McMeekin, 2011).

2.3.2 Institutional barriers

On the institutional level, scholars overall diagnosed an institutional lock-in associated with changing patterns of behaviour, social rules and norms that favour fossil-fuel-based technologies which have been deployed throughout the last decades. These can be categorised into hard (e.g. legal) and soft (e.g. norms) institutional problems (Chadha, 2011; Foxon & Pearson, 2008; Hekkert & Negro, 2009; Klein Woolthuis et al., 2005; Rennings, 2000). In addition information asymmetries (incomplete or imperfect information) translate into bounded rationality that prevents clean technologies from being developed and deployed (Jaffe et al., 2005; Jaffe & Stavins, 1994; Sanstad & Howarth, 1994; Schleich, 2009).

These abstract obstacles translate into tangible barriers along the innovation process. During the demonstration stage, infrastructure problems, including physical such as power and transport and scientific infrastructure such as high-quality universities, research laboratories and technical institutes represent a significant barrier since eco-innovation as systemic innovations depend on complementary, capital intensive assets for their commercialisation e.g. in the case of fuel cell mobility (Foxon & Pearson, 2008; Köhler et al., 2010; Steinbach, 2013; Zhang et al., 2012).

When moving towards commercialisation and introduction to markets, regulatory risk and uncertainty (Bergek et al., 2013; Blyth et al., 2007; Böhringer et al., 2009; Foxon et al., 2005; Haley & Schuler, 2011; Wüstenhagen & Menichetti, 2012) such as unanticipated or recurring policy changes, legal security and duration of administrative processes proves significantly hindering, even when the technology is fully mature as low-carbon innovation exhibit a high regulatory dependency (Lüthi & Prässler, 2011; Lüthi & Wüstenhagen, 2012).

Beginning with pre-commercial phase and deployment, local and environmental acceptance that includes technological, economic, administrative approval and spatial planning (Dinica, 2008; Iyer et al., 2013; Sovacool, 2009a; Steinbach, 2013), negative attitudes and social values or pressure from communities hinder the spreading of innovative clean technologies (Montalvo, 2008; Smink, Hekkert, & Negro, 2013; van den Bergh, 2013). Examples include the deployment of renewable energy (RE) or carbon capture and storage.

2.3.3 Economic barriers

Economic barriers represent significant obstacles to low-carbon innovation since these technologies are subject to economic lock-in and corresponding path dependency due to a history of investments in fossil-fuel based technologies which are thus economically more competitive (Foxon & Pearson, 2008; Rennings, 2000; Wüstenhagen & Menichetti, 2012). In addition, innovative clean technologies are subject to externalities since the prices for fossil-fuel-based technologies do not incorporate their negative environmental effects (Jaffe et al., 2005; Jaffe & Stavins, 1994).

In the basic and applied R&D phases (technology generation), these failures translate into private underinvestment in R&D (Baker, Chon, & Keisler, 2009; Mowery et al., 2010; Nemet & Kammen, 2007; Sovacool, 2008) as well as a general product and market uncertainty (Barreto & Kemp, 2008; Bosetti & Tavoni, 2009; Montalvo, 2008). From demonstration to supported commercial stages, scholars refer to high transaction costs associated with the introduction and use of new technologies (Sanstad & Howarth, 1994; Sorrell, 2007), costs for deployment (Bergek et al., 2013; Kimura, 2010; Kobos, Erickson, & Drennen, 2006), high discount rates on future savings ('energy technology paradox') and a corresponding 'waiting' for improvements as main barriers to commercialisation (Jaffe & Stavins, 1994; Schleich, 2009; van Soest & Bulte, 2001).

Throughout all phases of market introduction, even when the technology is fully commercialised, market organisation questions, such as the missing development of niches, hinder widespread adoption (Jacobsson & Lauber, 2006; Smink et al., 2013). A lack of business models for radically new clean technologies such as the quality and price of maintenance services proves to be equally challenging (Dinica, 2008; Kley, Lerch, & Dallinger, 2011).

In addition to these organisational questions, flawed pricing mechanisms occur due to the missing incorporation of real environmental benefits. Many clean technologies depend on energy savings which are foiled by artificially low energy prices due to subsidies for fossil-fuels (Jaffe et al., 2005; Jaffe & Stavins, 1994; Jefferson, 2008; Sandén & Azar, 2005; Sovacool, 2008). This translates into long timescales for turnover especially in energy-supply and energy end-use technologies and for development and demonstration of new energy technologies (Haley & Schuler, 2011; Kenney & Hargadon, 2012).

Throughout the niche-market and fully commercial phases, demand articulation failures occur, since there are insufficient spaces for anticipating and learning about user needs to enable the uptake of low-carbon innovations by users. The absence of orienting and stimulating signals from public demand and a lack of demand-articulating competencies in the private sector further aggravate the problem (Markard & Truffer, 2008; Weber & Rohracher, 2012).

2.3.4 Financial barriers

Overarching genuine financial barriers i.e. barriers that relate to financial markets consist of the information asymmetries and bounded rationality as financiers typically do not possess technological or political know-how to evaluate risks and returns of investments in innovative clean technologies (Olmos et al., 2012).

Genuine financial barriers to low-carbon innovation typically start with the demonstration and early commercialisation phase, when external finance is needed. Here, scholars diagnose capital market imperfections (Jacobsson & Karltorp, 2013; Leete et al., 2013; Sanstad & Howarth, 1994) for innovative clean technologies as venture capital (VC) is missing or is unsuitable for certain investments (Kenney & Hargadon, 2012; Leete et al., 2013; Randjelovic et al., 2003). Here price, volume and balancing risks are the main underlying financial barriers (Bergek et al., 2013; Mitchell, Bauknecht, & Connor, 2006). In that respect, scalability i.e. the seamless switch from demonstration scale to production scale, proves to be a challenge (J. Brown & Hendry, 2009; Hendry et al., 2010).

Further obstacles in the diffusion stages include slow capital stock turnover and a corresponding long payback period which relates to capital intensity, especially high upfront investments that hinder the ability to finance by institutional investors i.e. private and public equity or mezzanine capital. Similarly firms are unsuccessful in obtaining credit for investments (Iyer et al., 2013; Jacobsson & Karltorp, 2013; Leete et al., 2013; Lüthi & Prässler, 2011; Schleich, 2009; Zhang et al., 2012).

2.3.5 Political barriers

Political barriers directly relate to competencies and mandates of policy makers that engage in the innovation process for clean technologies. First, policy coordination failures occur. These include the lack of multi-level policy coordination across different systemic levels (e.g. regional–national–European or between technological systems), the lack of horizontal coordination between

STI policies and sectoral policies (e.g. transport, energy, agriculture) as well as the lack of vertical coordination between ministries and implementing agencies which leads to deviations between strategic intentions and operational policy implementation. At the intersection between public and private actors, scholars reveal missing coherence between public policies and private sector institutions. A lack of temporal coordination also results in mismatches related to the timing of policy interventions (Geels, 2010; Weber & Rohracher, 2012). Second, reflexivity failures occur which pertain to the insufficient ability of the system to monitor, anticipate changes and involve actors in processes of self-governance, as well as provide spaces for experimentation and learning. Correspondingly, policy makers do not implement adaptive policy portfolios to keep options open and deal with uncertainty (Geels, 2010; Weber & Rohracher, 2012). Third, Weber and Rochracher (2012) refer to a directional failure, which comprises a lack of shared vision regarding the goal and direction of the transformation process, the inability to coordinate distributed agents involved in shaping systemic change and insufficient regulation or standards to guide the direction of change.

These abstract policy barriers translate into adverse effects of policies in the early stages (Schmidt & Marschinski, 2009), as the fundamental decision between mission oriented vs. level playing field policy has not yet been resolved (Kern & Smith, 2008). In later commercialisation stages, inefficient allocation of planning and authorisation competencies have been highlighted (Friebe, Flotow, & Täube, 2014; Steinbach, 2013).

2.3.6 Interaction and capability barriers

Barriers to agents interacting along the innovation process for clean technologies include network problems and externalities. Unintended knowledge spillovers prevent firms from commercialising new inventions as they cannot harness the benefits from their research. The linkages between actors could be too weak or too strong which results into a blindness to what happens outside the network in the TIS (Iyer et al., 2013; Jaffe et al., 2005; Popp, 2010).

More concretely, missing stakeholder involvement proves to be a significant barrier since clean technologies usually affect a range of stakeholders throughout development, demonstration and especially diffusion phases e.g. for renewables or smart grids (Enzensberger, Wietschel, & Rentz, 2002; J. Hall & Kerr, 2003; Zhang et al., 2012).

Capability problems arise when inefficient firms especially small and medium-sized enterprises (SMEs) and research institutes possess limited organisational capabilities, that which hinder their capacity to adopt or produce new clean technologies over time (Klein Woolthuis et al., 2005; Montalvo, 2008). This translates into insufficient human capital for instance (Jacobsson & Karltorp, 2013).

2.3.7 Transformation barriers

Drawing from the literature on system transitions, scholars highlight transition problems. These problems prevent IS from transitioning towards a more sustainable path (Foxon & Pearson, 2008; Markard et al., 2012). Lock-in problems, derived from socio-technological inertia, that might hamper the emergence and dissemination of clean technologies are comprised of the aforementioned technological, economic and institutional lock-ins (Chadha, 2011; Foxon & Pearson, 2008; Rennings, 2000).

Beginning with the commercialisation and diffusion stages, Weber and Rorbacher (2012) add the category of complementary socio-technical barriers that impede the transition of the innovation system towards sustainability. These comprise behavioural and cultural barriers such as social interests of the incumbents (i.e. firms developing and applying fossil-fuel based technologies) and the legitimacy of the new technology thus creating regime dynamics which involve actors and institutions, especially in the demonstration stages (A. Smith et al., 2010; Sovacool, 2009a). Hockerts and Wüstenhagen (2010) state that the interaction between incumbents and new entrants provides the opportunity to transfer eco-innovation from niches into the mainstream markets. However, during these stages the power relations across the networks of actors involved in a regime typically prevent a systemic change of technologies, markets and institutions (Kern & Smith, 2008; A. Smith, Stirling, & Berkhout, 2005).

2.4 Policy approaches to address the barriers

Figure 2-3 illustrates policy levers at different stages in the innovation cycle to address the above mentioned barriers and tailor their intervention. Measures such as technology policies, encouraging interaction and learning should be focused on the early stages in the innovation cycle whereas capital market support and transformation-enabling policy instruments should be deployed at the demonstration stage onwards, when most of the economic, financial and institutional barriers interact. This period, known as the 'valley of death' is typically challenging for any kind of innovation, however in the light of the aforementioned complexity of the barriers particularly challenging. Hence, policy makers should have a holistic approach to market creation, thereby designing and adjusting their policy portfolio throughout the innovation cycle. As a result policy measures targeting the critical stages should start earlier in the innovation process to allow for an efficient transition between the phases. Institutional innovation ought to start early in the innovation cycle to accompany technology generation and fostering interaction. Support for learning should be extended towards the pre-commercial phases to address barriers related to acceptance.

Figure 2-3: Policy responses to accelerate commercialisation and diffusion of eco-innovation

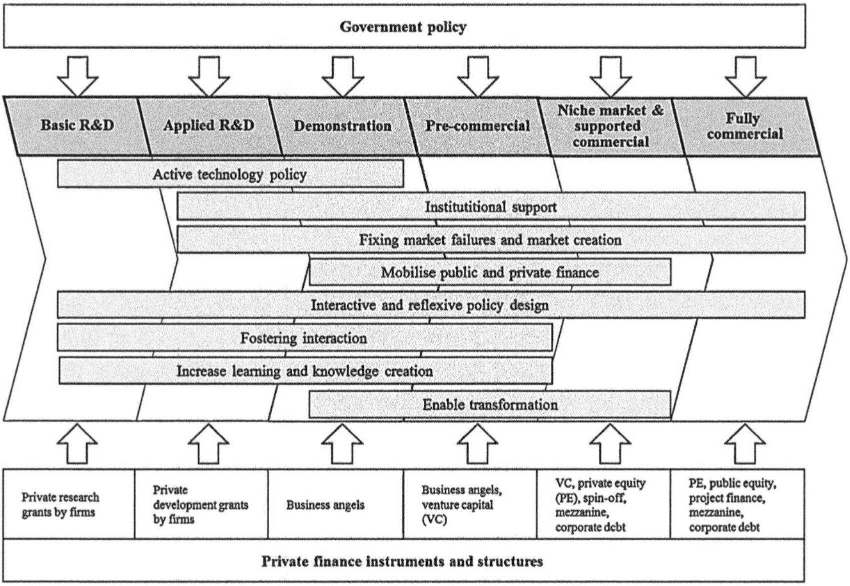

Own representation. Framework adapted from Auerswald & Branscomb (2003); Bürer & Wüstenhagen (2009); Wüstenhagen & Menichetti (2012).

2.4.1 Active technology policy

In order to address technological barriers such as technological lock-in and path-dependency as well as learning and knowledge deficits throughout the innovation cycle, scholars suggest a long-term technology strategy (Blanford, 2009). This refers to effective coordination with demand-side policies, support for transformational change across the research development and demonstration (RD&D) spectrum of activities, and focused efforts on identifying and attacking 'reverse salients' in existing technologies (Hargadon, 2010).

More concretely, to address the slow diffusion of knowledge and learning in the early stages, patent commons and the creation as well as support of technology networks are suggested (B. H. Hall & Helmers, 2011; Luiten & Blok, 2003; Luiten, van Lente, & Blok, 2006). Furthermore, to escape technological path-dependency, policy makers should aim at increasing technological diversity (Jefferson, 2008; Mowery et al., 2010; van den Bergh, 2013). This can be done by integrating environmental policy targets in technology policy and operationalising them into research programs (Kivimaa & Mickwitz, 2006).

During the critical demonstration and pre-commercial phases ('valley of death'), demonstration projects and trials (DTs) as well as technology transfer programs are strongly suggested to assess and validate feasibility and commercial viability (J. Brown & Hendry, 2009; Bürer & Wüstenhagen, 2009; Hendry et al., 2010; Kenney & Hargadon, 2012; Lewis & Wiser, 2007; Sartorius, 2008). In addition, these trials could be used to rule out emerging reverse salients.

2.4.2 Institutional support

Ecological and environmental economists widely suggest a combination of (environmental) regulation and R&D support (Fischer, 2008; Fischer & Newell, 2008; Leitner et al., 2010; Popp, 2010; van den Bergh, 2013). However, to escape institutional lock-ins and related failures, Rennings (2000) highlights the importance of systemic approaches to consider the variety of factors surrounding complex failures.

Therefore, policy makers aiming at accelerating commercialisation and diffusion should provide support for development of (grid-) infrastructure technologies and other complementary assets (Henriot, 2013; Jacobsson & Karltorp, 2013; Köhler et al., 2010). To mitigate problems associated with missing acceptance among the local population, an active stakeholder engagement proves to be effective starting with the applied R&D and demonstration phases already. Thus societal concerns need to be taken seriously and wherever possible, societal and local communities could be empowered (J. Hall & Kerr, 2003; Sine & Lee, 2009; Szarka, 2006).

To avoid regulatory risks or uncertainty in the diffusion stages, clear performance-based regulation such as eco-labelling and standards for energy audits and disclosure address corresponding barriers (Beerepoot & Beerepoot, 2007; Fischer & Newell, 2008; Jaffe & Stavins, 1994; Popp, Hafner, & Johnstone, 2011). Additionally, direct regulation for air, water, soil has proven to accelerate commercialisation and diffusion (Bañales-López & Norberg-Bohm, 2002; Bauman, Lee, & Seeley, 2008; Horbach et al., 2012; Perez, 2013; Saint Jean, 2008).

2.4.3 Fixing market failures and market creation

To address externalities as well as other reasons for economic path-dependencies, economists and innovation scholars theoretically agree that a combination of subsidies with taxes and regulation encourages eco-innovation (Loiter & Norberg-Bohm, 1999; Nesta, Vona, & Nicolli, 2014; Veugelers, 2012). Negative externalities such as CO_2 emissions should be internalised through a GHG-emission trading system (Fischer & Newell, 2008; Goulder & Mathai, 2000; Jakeman et al., 2004; Mathews et al., 2010; Popp, 2010; Rogge et al., 2011). Another possibility would be long-term policy signals such as GHG concentration targets or CO_2 emission reduction targets (Popp, Hascic, & Medhi, 2011). Although this is theoretically optimal, research shows that more policy instruments are needed to address economic barriers at different stages in the innovation cycle.

Above all, in an effective STI policy technology push mechanisms (R&D policies) complement demand pull deployment policies (Veugelers, 2012). To address under-investment in R&D in the early stages, R&D subsidies and grants (Acemoglu, Aghion, Bursztyn, & Hemous, 2012; Jacobsson & Lauber, 2006; Jaffe et al., 2005; Popp, 2010; Schilling & Esmundo, 2009) or R&D tax credits (Freeman, 1996; Jaffe & Stavins, 1994; Loiter & Norberg-Bohm, 1999) have been suggested to alleviate financial constraints and to increase spillovers (Kverndokk, Rosendahl, & Rutherford, 2004; Nemet, 2012). Targeting university R&D seems especially promising (Kenney & Hargadon, 2012). The more uncertain the technologies, the higher investments should be (Bosetti & Tavoni, 2009). However low-carbon innovation programs need to be shaped and evaluated as portfolios that distribute their investments across degrees of risk and time frames for anticipated returns (Jaffe et al., 2005). A complementary reduction of R&D subsidies for fossil-fuel-based technologies is also strongly suggested (Jefferson, 2008; Schilling & Esmundo, 2009; Sovacool, 2008). Beginning with the applied R&D phase to overcome the 'valley of death', in addition to continuous public investment in R&D and commercialisation (Kimura, 2010), production support measures, such as production tax credit, should be enacted (Barradale, 2010; Frondel, Horbach, & Rennings, 2007; Haley & Schuler, 2011; Komor & Bazilian, 2005).

Demand-pull policies consist of consumption support measures (Haley & Schuler, 2011; Montalvo, 2008; Sartorius, 2008). These could take the form of tax breaks and incentives for entrepreneurs to gain a competitive advantage vis-à-vis incumbents (Bürer & Wüstenhagen, 2009; Komor & Bazilian, 2005). As market formation becomes imperative in this stage, policy makers should connect market formation and policy incentives through neutral support as more market segments are targeted (del Río & Bleda, 2012; Dewald & Truffer, 2011). Lead market creation on the other hand is also a viable policy option (Beise & Rennings, 2004; Horbach, Chen, Rennings, & Vögele, 2013). To support the commercialisation of clean technologies throughout the demonstration phase, public procurement as a mission-oriented innovation policy could be used (e.g. Bürer & Wüstenhagen, 2009; Edler & Georghiou, 2007; Edquist & Zabala-Iturriagagoitia, 2012; Foxon et al., 2005; Freeman, 1996; Mowery et al., 2010).

Throughout the niche-market and fully commercial stages, subsidies (e.g. re-fund schemes) could accelerate the diffusion in the short run (Cantono & Silver-berg, 2009; Fischer & Newell, 2008; Komor & Bazilian, 2005; Montalvo, 2008). Apart from withdrawing subsidies for fossil fuel based technologies (Bürer & Wüstenhagen, 2009; Jaffe & Stavins, 1994; Jefferson, 2008; Sovacool, 2008), taxes on products, emissions or fossil fuels (Acemoglu et al., 2012; Fischer & Newell, 2008; Freeman, 1996; Mickwitz, Hyvättinen, & Kivimaa, 2008; Popp, 2010) or stable tax incentives for private innovation further stimulate competitiveness with fossil-fuel based technologies and thus encourage diffusion in the long-run (Kern & Smith, 2008).

Furthermore on the one hand, policy makers should focus on the product standards and demand-generating effects of regulation as well as an articula-tion of quality requirements (M. A. Brown, 2001; Bürer & Wüstenhagen, 2009; Freeman, 1996; Mickwitz et al., 2008; Oltra & Jean, 2005; Perez, 2013; Rennings & Rammer, 2011). On the other hand, certain industries require de-regulation (e.g. energy) for clean technologies to succeed in the market place (Kenney & Hargadon, 2012).

Finally feed-in tariffs (Bürer & Wüstenhagen, 2009; del Río & Bleda, 2012; Dinica, 2006; Johnstone et al., 2010; Lewis & Wiser, 2007; Mitchell et al., 2006) and renewable obligation certificates (ROC) or quota models such as renewable portfolio standards (RPS) (Carley, 2009; Lewis & Wiser, 2007; Mitchell et al., 2006) and green certificates have been proven to accelerate the diffusion of RE technologies (Bergek & Jacobsson, 2010; Bird, Holt, & Levenstein Carroll, 2008; Komor & Bazilian, 2005; Menanteau, Finon, & Lamy, 2003).

2.4.4 Mobilise public and private investment

Throughout the early R&D stages, Olmos et al. (2012) analyse which instruments maximise the amount of socially valuable clean RD&D by leveraging private sec-tor funding as far as possible within each stage of project maturity. They suggest public loans, or guarantees provided by public bodies backing private loans, along with public investments in the equity of innovating companies depending on the characteristics of the research projects.

Moving towards commercialisation, financial barriers to market creation could be mitigated by either directly investing into infrastructure and companies or by incentivising private investments into clean technologies. Thus, combined public and private investment and state investment banks represent vehicles of direct intervention (Mathews et al., 2010; Mazzucato, 2013). Improving positive expectations of future market opportunities, encouraging private capital into the less mature and difficult-to-finance technologies and the regulation of financial markets to redirect financial capital in productive investments represent incentives for financiers (Foxon & Pearson, 2008; Jefferson, 2008; Perez, 2013). Thus direct financing, investment enabling, and fiscal policies represent a powerful policy mix to address financial barriers (M. A. Brown, 2001; Foxon et al., 2005; Perez, 2013).

Specific measures during the commercialisation phase could include the creation of public-private-partnership (PPP) VC funds or statutory obligations, grants or capital-expenditure, and fiscal incentives such as tax breaks for investors (Bürer & Wüstenhagen, 2009; Foxon et al., 2005; Mathews et al., 2010). To support the fully commercial phase, governments should consider establishing PPP private equity funds to leverage investments in larger infrastructure or mature cleantech companies (Mathews et al., 2010).

2.4.5 Interactive and reflexive policy design

To address political barriers such as policy coordination failure, directionality failure and reflexivity failure, scholars suggest a number of overarching design features for low-carbon innovation policy. First of all, policy design should adhere to certain criteria such as flexibility, stability, targeting, stringency and predictability (Arent, Wise, & Gelman, 2011; Foxon & Pearson, 2008; Jacobsson & Karltorp, 2013; Leete et al., 2013; Lewis & Wiser, 2007; Mickwitz et al., 2008; Noailly & Batrakova, 2010; Rogge et al., 2011). Second, the timing of policy and inter-temporal consistency of the policy measures are important to achieve the desired results (Loiter & Norberg-Bohm, 1999; van den Bergh, 2013; Veugelers, 2012). Third, policy regimes should be evaluated according to outcome indicators of technology, actors and institutions as well as societal and environmental impact (Jaffe, Newell, & Stavins, 2002; Jaffe et al., 2005; Neij & Åstrand, 2006). This could be done using an interactive approach to policy design which targets market design and implications as well as stakeholder involvement (Enzensberger et al., 2002). In addition, policy makers should continuously check the validity of the models and projections of global climatic change, and audit these projections to ensure that planning systems allow questioning of commercial and technical criteria (Arent et al., 2011; Jefferson, 2008).

In addition to the overall recommendations, scholars suggest a portfolio of policy measures at different stages in the innovation cycle (Bürer & Wüstenhagen, 2009; Noailly & Batrakova, 2010). With regard to administrative (operational) barriers, a single authority planning, authorisation and regulation competencies may abolish the existing lack of coordination, and stricter administrative time-limits and sanctions could accelerate the development of complementary assets, such as infrastructure (Steinbach, 2013).

2.4.6 Fostering interaction, increasing knowledge and learning

The interaction barriers such as network externalities, spillovers and missing stakeholder involvement can be addressed by a number of different cooperative public-private arrangements. Above all, policy makers should encourage open innovation to increase the knowledge base and to increase technological spillovers from other countries (Mowery et al., 2010; Verdolini & Galeotti, 2011). To gain broader momentum for technology development and diffusion, policy makers work with members of different technology-specific advocacy coalitions, both private capital and various interest organisations, and involve social movements as well as stakeholders, especially for systemic innovations that require public acceptance (e.g. large scale energy infrastructure) (J. Hall & Kerr, 2003; Jacobsson & Bergek, 2004; Jacobsson & Lauber, 2006; Sine & Lee, 2009).

During the R&D stages of the innovation cycle, research has shown that strategic research partnerships (public–private RD&D partnerships) among various combinations of industry, academia, national laboratories, other governmental and non-governmental entities (NGOs) such as SBIR, ATP or ARPA-E[6] are vehicles to overcome cooperation barriers and competence lock-ins (M. A. Brown, 2001; Chadha, 2011; Freeman, 1996; Jaffe et al., 2005; e.g. Kenney & Hargadon, 2012; Sartorius, 2008; Sovacool, 2008). In addition, clusters for clean technologies could support increased knowledge spillovers (Barreto & Kemp, 2008). Beginning with the demonstration phase, scholars suggest the establishment of platforms for old and new technologies (A. Smith et al., 2010). During the diffusion stages, policy makers could facilitate knowledge networking through markets and otherwise for firms, universities and other organisations (Foxon & Pearson, 2008; Freeman, 1996).

---

6   These refer to technology programs in the US: Small Business Innovation Research (SBIR); Advanced Technology Program (ATP); Advanced Research Projects Agency – Energy (ARPA-E).

To address capability barriers such as inadequate human capital or inefficient firms, research highlights learning & knowledge integration and continued education (Sartorius, 2008; A. Smith et al., 2010). Thus in the early stages, training human capital in relevant areas is needed for commercialisation and diffusion (Jacobsson & Karltorp, 2013). Additionally, providing personnel, especially managers, with an understanding of the scale, unpredictability of costs, and the length of time required to develop technologies, is paramount (Leete et al., 2013). In the diffusion phase, voluntary, information and technical assistance programs embedded into campaigns could enhance the deployment of innovative clean technologies (M. A. Brown, 2001; Jaffe & Stavins, 1994; Newell et al., 2006).

2.4.7 Facilitate transformation

The literature on socio-technical transitions informs innovation scholars about possible ways to address transition barriers. Coenen and Diaz Lopez (2010) compare different approaches to system failures for eco-innovations and conclude that a combination of the focus on global economic competitiveness and a sustainable transition of society would be most fruitful.

To support these processes effectively, policy makers need to develop a range of characteristics – high analytical competence, in-depth knowledge of relevant technological systems, co-ordination skills, patience, flexibility and political strength. Additionally they are well advised to integrate knowledge development and management for sustainability into their policy making (Jacobsson & Bergek, 2004; Shin, Curtis, Huisingh, & Zwetsloot, 2008).

Throughout the critical phases of the innovation cycle and to overcome the gap between demonstration, pre-commercial and supported commercial phases, niche market creation and strategic niche management is suggested to challenge incumbents and regime technologies (Foxon & Pearson, 2008; Jacobsson & Lauber, 2006; Kern & Smith, 2008; Kimura, 2010; Smink et al., 2013; A. Smith et al., 2010). An integration of new technologies into existing systems facilitates this process (Arent et al., 2011). To address the socio-economic impediments to eco-innovation, social persuasion, demand factors, and constructive technology assessment could be valuable (Freeman, 1996).

2.5 Barriers to low-carbon innovation and policy responses

2.5.1 Discussion

The objectives of this literature review have been to firstly review high-quality published literature revolving around systemic or market failures for low-carbon innovation, secondly to organise it along the innovation process to connect abstract failures, tangible barriers and policy responses and thirdly to identify empirically under-researched areas.

The review analysed relevant literature streams that inform the innovation process of low-carbon innovation, namely innovation systems (IS) (Jacobsson & Bergek, 2011), transition studies (Markard et al., 2012), environmental and ecological economics (Fischer & Newell, 2008; Newell et al., 2006) as well as energy economics and policy (Jakeman et al., 2004; Popp, 2010). Each of those literature streams has unique contributions to the discussion[7]. First, the IS literature provides the system failures framework as well as a differentiated view on actors, networks and institutions and a policy orientation. Second, the transition studies literature stream highlights conditions and interactions integrated in the multi-level-perspective to identify barriers that prevent the innovation system for clean technologies over fossil-fuel based innovation systems. Third, the environmental and ecological economics literature streams provide integral insights into innovation in clean technologies, particularly market based-mechanisms, policy instruments and models for technological change. Fourth, as a large part of clean technologies and eco-innovation unfolds in the energy sector, insights into technological specificities as well as dynamics in this highly regulated sector with a focus on policy recommendations is necessary.

By integrating these streams of literature in a theoretical framework that models the innovation process for clean technologies (thus adding a temporal perspective) and organising the barriers and policy responses according to a failures or problem-oriented IS approach, this article provides the reader with a systemic view to connect abstract failures that are often subject to academic research in economic models, to specific barriers along the innovation chain as put forward by scholars of the innovation system and energy policy literature. This link connecting concrete innovations, regulation and the behaviour of actors has been missing throughout the literature (Dewald & Truffer, 2011; Newell et al., 2006; van den Bergh, 2013).

---

7   The most influential journals in the field comprise: *Innovation systems* – Research Policy, Technological Forecasting and Social Change, Industrial and Corporate Change, Industry & Innovation; Technovation; *Transition studies* – Environmental Innovation and Societal Transitions, *Environmental and ecological economics* – Journal of Environmental Economics and Management, Ecological Economics, Environmental and Resource Economics, *Energy* – Energy Economics, Energy Policy, *Interdisciplinary* – Journal of Cleaner Production.

Environmental, ecological and energy economics examine the mechanisms of different policy instruments on technological change to overcome abstract market-related barriers by applying empirical and conceptual models (Fischer, 2008; Popp, 2006). Whereas this gives guidance to the overall effectiveness of the instruments, implementation of the instruments as well as boundary conditions determine the actual outcomes as well as the efficiency (Johnstone et al., 2010; Kley et al., 2011; Neij & Åstrand, 2006; Rogge et al., 2011; Smink et al., 2013). In this regard the innovation systems and energy policy literatures provide the essential context-specific insights by using qualitative case studies and quantitative empirical analyses. Additionally socio-economic factors, which have been subject to analysis in the transition literature further impact the rate and direction of technological change and innovation towards sustainability (Jacobsson & Bergek, 2011; Jefferson, 2008; Kern & Smith, 2008). Providing policy makers with an understanding of abstract market failures, tangible technology case studies and boundary conditions enables them to effectively and efficiently support low-carbon innovation.

A combination of factors such as financial, economic, institutional and transition barriers slows down the development, commercialisation and diffusion of clean technologies and the overall technological transformation (Iyer et al., 2013). This in turn inhibits the politically-induced transition processes. Advancing systems thinking in the field of eco-innovation as argued by Jacobsson & Bergek (2011) and Foxon & Pearson (2008) is crucial to mitigate these failures. Moreover this article confirms and concretises earlier work which argues that gaps exist in the transition from the demonstration stage towards the pre-commercialisation stage, and between the pre-commercialisation and supported commercialisation stage (scaling) due to technological, market, regulatory, systemic risks (Foxon et al., 2005, 2008; Foxon & Pearson, 2008).

Scholars started to explore systemic instruments and combinations of instruments to address this complex web of barriers, e.g. combinations of technology-push and demand-pull mechanisms with environmental regulation measures to achieve a level-playing field with fossil-fuel based technologies (Foxon & Pearson, 2008; Hargadon, 2010; Mowery et al., 2010). However, this research neglected the time-dimension, which this paper integrates by developing an innovation cycle framework. Policy makers need to tailor their responses according to the stage in the innovation cycle and the problems occurring at this specific stage (see Figure 2-2 and Figure 2-3).

During the technology generation phase i.e. basic and applied R&D stages, technological, *interaction and capability barriers* are most salient. Here, policy makers can rely on a number of instruments, including *active technology policy, fostering interaction and increase learning and knowledge creation.* Throughout the demonstration and pre-commercial phases, low-carbon innovations are subject to a complex web of barriers, especially regarding *market development, corresponding institutions and financial resource allocation.* Policy makers could tackle these inefficiencies especially by *institutional support, fixing market failures and market creation, and corresponding mobilisation of public and private finance.* These elements should enable the *transition towards a sustainable innovation system.* During the later stages of the eco-innovation technology lifecycle i.e. niche-market and supported commercial as well as fully commercial stages, most of the barriers to commercialisation prevail. A combination of *addressing market failures, market creation and institutional support* proves most useful in the later stages. Summing up, the research undertaken offers policy makers a menu of options to address barriers to eco-innovation, and is therefore highly relevant for the ongoing policy debate and a concrete operationalization for the 'entrepreneurial state' (Mazzucato, 2013).

2.5.2 Avenues for future research

The comprehensive integration of literature streams provides a variety of avenues for future research. For example on a conceptual level, the temporal perspective could be integrated more holistically in the concept of TIS. In addition the failure categorisation could be used to identify systemic problems in sectors similar to clean technologies, such as pharma or biotechnologies. The review further gives preliminary indications that different barriers impact the financing of eco-innovation. Thus especially the underlying technological, institutional, political, economic and transformation barriers have consequences for the finance environment.

Throughout this chapter, the interplay between finance and innovation focusing on public-private interaction and policies is highlighted. This step has been taken for a number of reasons. First, financial barriers to low-carbon innovation have been discussed mostly on a firm level, neglecting the technological level (Este, Iammarino, Savona, & Tunzelmann, 2012; Pellegrino & Savona, 2013). Second, the transformational aspect of finance with regard to innovation processes has been under-researched. Financiers play a special role in transitions as they accumulate the necessary resources for large scale investments (Hargadon, 2010; Kenney & Hargadon, 2012; Perez, 2013). Third, the overall link between finance and innovation has been neglected throughout the TIS literature (Dosi, 1990; O'Sullivan, 2006; Perez, 2002). Table 2-1 shows possible avenues for future research in the domain of financing eco-innovation and adjacent areas.

Table 2-1: Selected avenues for future research

Area of study	Under-researched aspects
Financing eco-innovation	Advanced private financing options (beyond) VC for the commercialisation and diffusion phases Investor behaviour regarding cleantech Financial regulation and support for cleantech
Structures for public-private cooperation to increase eco-innovative activity	Micro-level of interaction between public and private actors in cleantech partnerships Determinants of these interactions in the diffusion stages Possibilities and limits for public-private cooperation
Effective and efficient policies for eco-innovation	Combination of support and financial instruments Comparison of policy instruments to support the diffusion Transformational policy instruments (governance)

Financing eco-innovation

A complex set of barriers revolves around the question of how to finance companies, projects and infrastructure based on eco-innovation. Financial barriers result mainly from economic barriers, such as failing markets. However a combination of technological barriers combined with institutional, interaction, capability and political barriers contribute to companies, projects and infrastructure failing to obtain finance. Most of the studies analysing the relationship between finance and innovation (with a few exceptions), focus on the generation and commercialisation of technologies, mainly highlighting equity and particularly VC as a suitable solution for certain types of companies and technologies (Bürer & Wüstenhagen, 2009; Kenney & Hargadon, 2012; Leete et al., 2013). However, a holistic approach taking into account the boundary conditions for innovations in the later stages (e.g. with regard to project finance or infrastructure investments) is still missing. Depending on the actual step in the innovation cycle, financing solutions for innovative companies and complementary infrastructure exhibit different characteristics (Henriot, 2013; Jacobsson & Karltorp, 2013; Leete et al., 2013; Mathews et al., 2010).

Initial research has been done to consider financing in the early stages beyond basic research grants such as public investments or loans, however public and private instruments in the transition towards demonstration, pre-commercial and supported commercial phases have not been analysed yet (Dinica, 2008; Mathews et al., 2010; Mazzucato, 2013; Olmos et al., 2012). Mechanisms to leverage the financiers capabilities are also lacking (Foxon & Pearson, 2008; Jefferson, 2008). For instance, to share the risks associated with early stage technologies and the tendency of private VC to avoid the early stages, especially for low-carbon innovation, PPP-VC might be a viable structure (Mathews et al., 2010). Initial analysis of the (non-financial) barriers suggest that these interact and result in consequences for financing (Bergek et al., 2013; Friebe et al., 2014; Wüstenhagen & Menichetti, 2012). Accordingly scholars call for an advanced consideration of risks and policy instruments that adjust the risk/reward ratio to successfully commercialise clean technologies (Foxon et al., 2008; Lüthi & Prässler, 2011).

The review therefore calls for an integrated approach on financing innovation, investigating concrete situations to overcome specific financing constraints. This includes looking at the financial regulation environment which indirectly impacts sources of capital available for clean tech innovation (Perez, 2013). All streams of innovation studies lack the specific relation with finance, respectively treat it as an input market in the IS or exogenous variable in the economics branches. The underlying reason might be a different logic of innovation and finance literatures (Dosi, 1990; O'Sullivan, 2006).

Structures for public-private cooperation to increase eco-innovative activity
Clean technologies face a number of barriers that relate to interaction of public and private actors, such as technological, capability and interaction barriers in the early stages as well as institutional barriers and economic and financial barriers in the later stages. To address both technological as well as interaction and capability barriers, strategic research partnerships among various combinations of industry, academia, national laboratories, other governmental and semi-governmental entities, and NGOs are suggested (M. A. Brown, 2001; Kenney & Hargadon, 2012; Sartorius, 2008; Sovacool, 2008). However, policy responses have so far neglected public-private cooperation and corresponding structures beyond the R&D stages in addressing the complex failures arising from public-private interaction such as transformational barriers or financial barriers. The determinants of the interactions as well as opportunities and limits for public-private cooperation need to be further explored.

Effective and efficient policies for eco-innovation
To address barriers to low-carbon innovation, policy makers are presented with a number of instruments along the innovation cycle. Prior research suggests that

traditional approaches to stimulating innovative activity could be amended by an open industry knowledge base for rapid diffusion of these technologies, thereby catalysing demand. A support for demonstration projects and cooperative public-private R&D programs should be combined with a significant share of private R&D investments to address the finance-related peculiarities of eco-innovation (Hargadon, 2010; Mowery et al., 2010). However, combinations of instruments have not been evaluated on a system level or with concrete relation to technologies.

Especially the demand-side measures should be evaluated with regard to their effectiveness, efficiency and consequences for finance to achieve synergies between public and private instruments (Haley & Schuler, 2011; Sartorius, 2008). This holds for instruments such as public procurement (Bürer & Wüstenhagen, 2009; Edquist & Zabala-Iturriagagoitia, 2012; Guerzoni & Raiteri, 2014; Mowery et al., 2010) or more holistic, lead market creation (Beise & Rennings, 2004; Horbach et al., 2013). A second avenue of research addresses investment-enabling and fiscal policies which have been neglected in current literature (M. A. Brown, 2001; Foxon et al., 2005; Perez, 2013). This relates to the overall research gap with regard to financing low-carbon innovation.

In sum, this literature review identifies a number of policy responses for barriers to low-carbon innovation along the innovation cycle, ranging from cooperation to regulation and towards support. To account for the systemic character of many barriers, firstly a combination of support and financial instruments could be analysed. Secondly a comparison of policy instruments regarding effectiveness and efficiency to support the commercialisation and diffusion is helpful in understanding interactions between barriers and policy instruments. These insights should help in conceptualising a policy mix that enables and governs the transition towards eco-innovation. A concrete application of this concept could be an analysis of the effects of policy instruments on financing for companies, infrastructure and projects.

2.6 Conclusion and policy implications

This article reviews the relationship between barriers to low-carbon innovation and policy solutions. The barriers are organised along the innovation cycle to provide policy makers with the ability to systematically and holistically treat the development, commercialisation and diffusion of clean technologies. Following from the above reviewed literature, eco-innovations are subject to both market and system failures. Thus a synthesis on the evidence from both TIS and innovation chain theory perspectives proves fruitful for the barriers and possible solutions. The abstract failures and policy responses translate into precise barriers and a portfolio of policy measures along the innovation chain. These should be designed to fit the actual technology and market lifecycles. The analysis is an attempt to create a basis for evidence-based policy to support policy makers in their decisions (Martin, 2012a).

Following the analysis, this article proposes an adaptive policy design to address specific barriers to low-carbon innovation along the innovation cycle for the technology which includes anticipating future steps in the technology development and commercialisation process and having the corresponding policy instruments ready to support a seamless transition between the stages and gaps. This requires strong signals from public actors towards research, industry and financiers (Mazzucato, 2013). In order to apply the multitude of policy instruments upon a complex web of barriers to clean technology innovation, policy makers need to develop the necessary skills such as in-depth knowledge of relevant technological systems, co-ordination skills, patience, flexibility (Arent et al., 2011; Jefferson, 2008; Veugelers, 2012). In sum, policy makers need to incorporate the 'adequate' distribution of responsibilities between private and public actors into their decision making.

The review gives preliminary indications that different barriers have consequences for the financing of eco-innovation, a link that has been neglected throughout the innovation studies community. Thus to accelerate the commercialisation and diffusion of clean technologies, policy makers need to address the genuine financial barriers (i.e. related to capital markets) but also the underlying technological, institutional, political and economic barriers as well as interaction, capability and transformation barriers that have consequences for the finance environment.

The coordination of policies for eco-innovation, especially STI policy and industrial policy and financial policy generates positive feedback loops and reaches a sustainable path for innovation (cf. Migendt, Schock, Flotow, Täube, & Polzin, 2014). In this respect, private finance plays a catalytic role. These actors provide the necessary investments into companies, projects and infrastructure, thus their view on the innovation process for eco-innovation and potential markets is crucial. Hence, the incorporation of risk/return perceptions throughout the policy design for different applications (companies, projects, infrastructure) along the innovation process is also vital.

Although this literature reviews aims to be transparent and replicable, there remain some limitations to the methodology used. While the databases do not contain all the relevant studies, they have nevertheless allowed building a sample that is representative of the work throughout the selected literature streams. This article focuses on implications of barriers to low-carbon innovation for financiers which, according to several scholars, represents a promising line of research (Jacobsson & Karltorp, 2013; Kenney & Hargadon, 2012; Mathews et al., 2010; Wüstenhagen & Menichetti, 2012). However more avenues of future research exist based on the analysis of this heterogeneous body of literature.

Acknowledgements
The author is grateful for the time and support of Paschen von Flotow (Sustainable Business Institute – SBI), Florian Täube (Solvay School of Management and Economics), Maria Savona and Paul Nightingale (SPRU – University of Sussex), Ruhi Deol (LMU Munich) as well as Michael Migendt (EBS Business School) for valuable comments on earlier versions of the paper. The researcher would like to thank the German Federal Ministry of Education and Research (BMBF) for their financial support as part of the research project "Climate Change, Financial Markets and Innovation (CFI)".

3 Accelerating the cleantech revolution: Exploring the financial mobilisation functions of institutional innovation intermediaries

Friedemann Polzin, Paschen von Flotow & Laurens Klerkx

Abstract: This research article explores the role of innovation intermediaries to accelerate the commercialisation of (clean) technologies. Drawing from the finance and innovation intermediaries literatures we show that (financial) barriers to (low-carbon) innovation can be partly overcome by particular functions of innovation intermediaries which in turn mobilises private finance along the innovation process. Therefore, we empirically evaluate roles and instruments of institutional innovation intermediaries (innovation intermediation, policy support, public-private cooperation, financial instruments). We contribute an intersection of the finance and innovation systems literature, by exploring the 'financial mobilisation functions' of innovation intermediaries to address barriers for (low-carbon) innovation along the innovation process.

3.1 Introduction

The on-going debate about how to mitigate climate change has encouraged policy-makers to conduct R&D for low-carbon innovations. The aim of these initiatives is twofold: firstly to reduce carbon emissions and secondly to foster long-term economic 'green' growth (OECD, 2009; Strand & Toman, 2010). However, complex system failures surrounding the commercialisation of eco-innovations[8] due to high uncertainty, missing carbon markets and resulting technological lock-in (Leitner et al. 2010).

Many (small and medium-sized) firms and research institutes invent technologies that are eventually not introduced to the market because of underinvestment in R&D or other (finance-related) market failures such as imperfect capital markets, scalability, asset intensity, missing complementary assets such

---

8   Based on previous literature (Foxon & Pearson, 2008; Horbach, Rammer, & Rennings, 2012; Rennings, 2000) this article adopts the following definition: Low-carbon innovation can be defined as the 'invention, commercialisation and diffusion of technologies that reduce carbon emissions and/or other environmentally negative impacts and thus contributes to sustainability'.

as infrastructure and an inadequate regulatory environment (Haley & Schuler, 2011; Kenney & Hargadon, 2012; Marcus et al., 2013; Mathews et al., 2010; Olmos et al., 2012). The incorporation of the 'finance perspective' at an early stage including a cooperation of innovative firms and research institutes with financiers could leverage public and private funds more effectively, enhance innovative activity and finally accelerate the commercialisation and diffusion process. Consequently especially for climate change-related eco-innovation, there is huge potential in connecting public support with private finance, as information asymmetries between innovators and financiers persist (Mowery et al. 2010).

Key actors in the innovation process include institutional (i.e. government affiliated) intermediaries that play a crucial role in establishing and governing a closer collaboration and in fostering knowledge flows between innovators and financiers to reduce information asymmetries and uncertainty (Hoppe & Ozdenoren, 2005; Howells, 2006; Kivimaa, 2014). While in recent years a lot of work has been done on innovation intermediaries (Howells, 2006; Katzy, Turgut, Holzmann, & Sailer, 2013; Klerkx, Álvarez, & Campusano, 2014; Klerkx & Leeuwis, 2009; van Lente, Hekkert, Smits, & van Waveren, 2003; Yusuf, 2008) resulting in conceptual and qualitative evidence that institutional intermediaries at the intersection of public and private R&D and commercialisation have beneficial effects (Kivimaa, 2014; Klerkx & Leeuwis, 2009; van Lente et al., 2003; Yusuf, 2008), specific research on the role of such organisations and individuals with regard to their financial innovation intermediation functions to address barriers in the innovation system remains scarce. These functions which could enable the transitions by integrating public policy and private finance have been largely neglected. We attempt to close this research gap, by addressing the following research question: *How do institutional innovation intermediaries address the complex set of barriers surrounding (low-carbon) innovation from R&D to commercialisation?*

While we address this question in the context of low-carbon innovation, as innovation system problems such as thin markets for finance, information asymmetries, failing markets for technologies are more pronounced there, we also belief it to be of relevance for innovation in general. We present qualitative in-depth evidence, exploring the financial mobilisation functions and the role of institutional intermediaries. The article is structured as follows: Section 3.2 outlines the theoretical underpinnings and integrates the streams of literature on innovation finance and innovation intermediaries. Section 3.3 sketches the methodological approach taken to assess the role of intermediaries and to evaluate their financial mobilisation functions. Section 3.4 presents the results, while section 3.5 mirrors these results to theory to draw conclusions (section 3.6).

3.2 Theoretical background

3.2.1 Financing of R&D and innovation

As regards financing innovation, scholars consider financiers as crucial to support the commercialisation and diffusion of new (clean) technologies (Hekkert & Negro, 2009; Hekkert et al., 2007; Perez, 2002; Schumpeter, 1939), and several researchers have pointed towards an underinvestment in R&D as a market failure for innovative activity in the early stages (B. H. Hall, 2002; B. H. Hall & Lerner, 2010; Myers & Majluf, 1984): Firstly, the market logic does not permit financiers to evaluate the quality of the research due to its highly uncertain nature (Akerlof, 1970; Arrow, 1962; Jaffe et al., 2005). Possible gains from R&D cannot be fully appropriated by the firm due to knowledge spillovers, i.e. the social returns are higher than the private return appropriated (Griliches, 1992; Jaffe et al., 2005). Secondly, imperfections in capital markets concern the fundraising capability of firms (B. H. Hall, 2002). While financing innovation and its related market failure are clearly an issue within the framework of innovation systems, the broad question of financing innovative activity has not been treated holistically, though several authors have indicated that the 'financial innovation system' underlying the national and technological innovation systems is a significant driver of innovative activity and should therefore include well-coordinated policies (Dahlstrand & Cetindamar, 2000; O'Sullivan, 2006; Perez, 2013; Wonglimpiyarat, 2011). Especially in the light of a transition towards clean innovation, private finance is highlighted as a critical factor (Leete et al., 2013; Mathews et al., 2010; Perez, 2013).

Within the innovation policy mix to enable this transition, different policy instruments are implemented (see Borras & Equist (2013) for an overview), of which economic transfers comprising different forms of finance is one of the categories. Different phases of the innovation process i.e. basic and applied R&D, demonstration and commercialisation, pre-commercial phases, niche-market and supported commercial as well as the fully commercial phase call for different forms of finance, the so-called finance chain of innovation (Auerswald & Branscomb, 2003).

In the basic and applied R&D phases, governments address underinvestment in risky R&D due to intangibility and limited appropriability with subsidies and grants (Dahlstrand & Cetindamar, 2000; Link & Scott, 2010). Moving to the commercialisation phases (Demonstration, pre-commercial, niche-market and supported commercial), as 'investment readiness' is proven by signalling quality of the business proposition linked to the emerging technology, external financiers such as business angels and venture capitalists (VCs) start financing (Mason &

Harrison, 2001). Informed financiers (i.e. so-called 'competent' VCs – Dahlstrand & Cetindamar, 2000) try to overcome underlying information asymmetries and other barriers, such as missing managerial talent, marketing capabilities or networks thereby reducing the monitoring and moral hazard problems (Da Rin, Nicodano, & Sembenelli, 2006; Holmstrom & Tirole, 1997; Repullo & Suarez, 2000). However, VCs have several shortcomings such as the necessity of a well-functioning equity market and a focus upon only certain industries at a time, which makes them unsuitable for investing in infrastructure, larger R&D projects or asset-heavy firms and projects (B. H. Hall, 2002; B. H. Hall & Lerner, 2010; Oakey, 2003). In addition, private equity, mezzanine and bank finance are often not available due to missing collateral or the overall level of risk related to the technologies and institutional environment (Ughetto, 2007, 2010). More mature firms often rely on internal funds however as commercial viability is often uncertain, the companies refrain from commercialisation activities. In many cases, this leaves structural holes (e.g. known as 'valley of death') in the commercialisation phase, since private equity, many VCs and credit financiers are often unable to seamlessly invest in companies that reach the end of the public R&D support phase or in complementary assets such as infrastructure required for commercialisation (Auerswald & Branscomb, 2003). In consequence this might lead to thin financial markets as difficulties arise in the supply and demand of finance. Simply increasing demand or supply is not sufficient as coordination problems often arise between innovators (e.g. entrepreneurs), financiers and government (Dahlstrand & Cetindamar, 2000; Nightingale et al., 2009). Policy makers could therefore systematically strengthen the market demand side by establishing public procurement programs or public private-research partnerships in order strengthen technological capability to support the supply side (Audretsch & Lehmann, 2004; Auerswald & Branscomb, 2003; Edquist & Zabala-Iturriagagoitia, 2012; Hargadon, 2010; Link & Scott, 2010).

In later stages of the innovation cycle (supported commercial and fully commercial), (clean) technologies face regulatory risks, flawed market pricing mechanisms or policy coordination failures (Foxon et al., 2005; Haley & Schuler, 2011; Weber & Rohracher, 2012). Hence governments could provide incentives to the financial sector and play a catalytic role in providing risk capital. This could be done by regulating certain industries, setting up institutions to make investments more profitable (Borrás & Edquist, 2013; Wonglimpiyarat, 2011) or by using direct instruments such as public procurement for innovation (Edler & Georghiou, 2007; Edquist & Zabala-Iturriagagoitia, 2012; Guerzoni & Raiteri, 2014). An overview of instruments used to finance innovation can be derived from Table 3-1.

Table 3-1: Overview of barriers to innovation and financing instruments

Phase in the innovation cycle	Barriers	Instrument	References
Basic and applied R&D	Intangibility and limited appropriability (Knowledge spillovers)	Subsidies, grants	(Kleer, 2010; Meuleman & De Maeseneire, 2012)
	Underinvestment in R&D	Tax credits	(Czarnitzki, Hanel, & Rosa, 2011; B. H. Hall & Lerner, 2010)
Demonstration and pre-commercial phase	Capital intensity Scalability	Mobilise private finance (Business angels, VC)	(Hendry et al., 2010; Kenney & Hargadon, 2012)
	Economic/ Technological/ Institutional lock-in	STI policy Regulation	(Foxon & Pearson, 2008; Klein Woolthuis et al., 2005; Rennings, 2000)
	Infrastructure	Public procurement Public-private-partnerships (PPP)	(Foxon & Pearson, 2008; Köhler et al., 2010)
	Market/demand articulation	Public procurement Effective coordination of demand-side policies	(Edler & Georghiou, 2007; Edquist & Zabala-Iturriagagoitia, 2012; Hargadon, 2010)
Niche-market and supported commercial	Missing VC Flawed pricing mechanisms	Strategic research partnerships (e.g. SBIR, ATP, ARPA-E)	(Audretsch & Lehmann, 2004; Auerswald & Branscomb, 2003; Link & Scott, 2010)
Fully commercial	Regulatory risks Policy coordination and reflexivity failures	Mobilise private finance (Private equity, banks, mezzanine, project finance)	(Foxon et al., 2005; Haley & Schuler, 2011; Weber & Rohracher, 2012)

3.2.2 The role of intermediaries in addressing (financial) barriers

One way to address the barriers and structural financial gaps in the innovation cycle (see Table 3-1) is having intermediaries between different actors (Howells, 2006). These actors intermediate knowledge, technologies and finance which is crucial for advancing markets (Boon, Moors, Kuhlmann, & Smits, 2008; Hoppe & Ozdenoren, 2005; Howells, 2006; Stewart & Hyysalo, 2008). Howells (2006,

p. 720) defines an innovation intermediary as *'an organisation or body that acts an agent or broker in any aspect of the innovation process between two or more parties. Such intermediary activities include: helping to provide information about potential collaborators; brokering a transaction between two or more parties; acting as a mediator, or go-between, bodies or organisations that are already collaborating; and helping find advice, funding and support for the innovation outcomes of such collaborations'*. Throughout this paper we focus on the functions of innovation intermediaries that relate to finance, since the mobilisation of financial resources is considered a key function of innovation systems which often hinders technologies from being developed and deployed (Bergek et al., 2008; Jacobsson & Bergek, 2011; Jacobsson & Karltorp, 2013). This function is further differentiated along the innovation cycle (see Table 3-2) as innovation intermediaries may adopt different roles along the phases for market creation (Howells, 2006; van Lente et al., 2003).

Table 3-2: Overview of policy instruments for financing innovation (implemented by intermediaries)

Category	Instrument	Corresponding phase in the innovation cycle	References
Direct	Subsidies, grants	Basic and applied R&D Demonstration and pre-commercial phase Niche-market and supported commercial	(Howells, 2006; Kivimaa, 2014)
	Tax credits	Basic and applied R&D	(Howells, 2006)
	Public procurement or Production support measures	Demonstration and pre-commercial phase Niche-market and supported commercial	(Kivimaa, 2014)
	Mobilise private Finance (Business angels, VC) / Improving positive expectations of future market opportunities	Demonstration and pre-commercial phase Niche-market and supported commercial	Highlighted as important by several authors (e.g. Howells, 2006; Kivimaa, 2014; Yusuf, 2008) *but underresearched*
Indirect	Support for regulation	Niche-market and supported commercial Fully commercial	(Kivimaa, 2014; van Lente et al., 2003)
	Support for science, technology and innovation (STI) policy	Niche-market and supported commercial Fully commercial	(Klerkx & Leeuwis, 2009; van Lente et al., 2003)

Category	Instrument	Corresponding phase in the innovation cycle	References
	Strategic research partnerships (e.g. SBIR, ATP, ARPA-E)	Basic and applied R&D Demonstration and pre-commercial phase	(Kivimaa, 2014; Klerkx & Leeuwis, 2009; Yusuf, 2008)
	Mobilise private Finance (Private equity, banks, mezzanine) / Improving positive expectations of future market opportunities	Niche-market and supported commercial Fully commercial	Highlighted as important by several authors (e.g. Howells, 2006; Kivimaa, 2014; Yusuf, 2008) *but underresearched*

During the basic and applied R&D phases which exhibit high technological and market uncertainty and a general underinvestment in R&D, innovation intermediaries help in finding new sources of capital for researchers, such as R&D programs, research grants and subsidies. Especially in public-private-partnership constellations that deal with the development of complex innovations with a highly uncertain outcome such as pharmaceutical or medical products which have parallels to clean technologies, a more active management approach is needed i.e. enhanced research governance in the form of bringing the necessary resources and stakeholders together and reducing development time and costs (Yaqub & Nightingale, 2012). Prior research also points out the selection of the most suitable finance mechanisms for each type of project (e.g. subsidy, revolving fund, loan, etc.) as equally relevant (Eickelpasch & Fritsch, 2005). The goals of the research efforts need to be aligned with the selection process of supported firms and corresponding financial support mechanisms (Santamaría, Barge-Gil, & Modrego, 2010). In this regard intermediaries might also be capable of sending signals to certify the quality of research (Howells, 2006; Yusuf, 2008).

During demonstration and pre-commercial phases where clean technologies exhibit high capital intensity and challenging scalability as well as missing complementary assets such as infrastructure and demand articulation problems, the distribution of R&D grants and demonstration support characterises resource allocation of innovation intermediaries (J. Brown & Hendry, 2009; Samila & Sorenson, 2010). Additionally they might coordinate public-procurement programs in order to increase demand. Beyond, innovation intermediaries may also engage into public-private-partnerships (PPP) between private financiers, government agencies and inventors or start-ups as seen in the SBIR and ATP programs[9]. The evaluation of these PPPs revealed mixed evidence with regard to commercialisation success of the participating firms (Audretsch et al., 2002; Chang, Shipp, & Wang, 2002; Lerner, 1999; Link & Scott, 2010). Clearly, these PPP programs address the underinvestment in R&D and commercialisation showing firms need financial support to scale up their operations when they have passed the seed and invention stage (Cooper, 2003). Put alternatively, government or government-affiliated entities such as intermediaries thicken up 'thin' financial markets for early stage innovations (Link & Scott, 2010; Mazzucato, 2013).

In the niche-market, supported commercial and the fully commercial phases the role of innovation intermediaries is less visible as the technology matures. However regulatory risks and the provision of complementary assets need to be handled to fully deploy technologies (Howells, 2006; Yusuf, 2008). Thus a relevant function during commercialisation and diffusion is the mitigation of uncertainty and risk between firms or research institutes and potential financiers as the latter are neither able to access the potential markets for the application of the novel technologies nor their surrounding institutional environment (regulation and STI policy). Intermediaries thus evaluate commercial value, reduce uncertainty and make the technology process more transparent (Hoppe & Ozdenoren, 2005).

Kivimaa (2014) thereby highlights the role of government-affiliated or institutional intermediaries to address (systemic) failures along the innovation cycle. These would be ideally positioned for this, because they intermediate knowledge and finance between public (research organisations, government) and private actors such as firms in the innovation system, maintaining a neutral and impartial position (Klerkx & Leeuwis, 2009). Institutional intermediaries are affiliated with policy bodies (i.e. ministries) whose aim is to advance technologies but also to foster markets and generate subsequent private investments after the R&D support phase (Yusuf, 2008). However, because of their affiliation with policy bodies, institutional intermediaries might be subject to limitations regarding neutrality and impartiality as conflicts of interests between the public bodies to which they are affiliated and private entities could arise (Klerkx & Leeuwis, 2009).

---

9   ARPA-E – Advanced Research Projects Agency – Energy, SBIR – Small Business Innovation Research, ATPAdvanced Technology Program refer to cooperative innovation programs in the US.

3.2.3 Research question and research gap

By drawing together previously separated literature streams of financing innovation and innovation intermediaries it becomes apparent that these agents could potentially play an important role in addressing (financial) barriers to (low-carbon) innovation along the innovation cycle as they hold a critical position between market actors and government. Previous research has looked at their functions (Hoppe & Ozdenoren, 2005; Howells, 2006), at user-producer interactions and demand articulation (Boon et al., 2008; Boon, Moors, Kuhlmann, & Smits, 2011), their role in commercialising research (Yusuf, 2008), their interaction with the policy environment (Klerkx & Leeuwis, 2009) and at their role with regard to transition towards sustainability in a broader frame (Kivimaa, 2014; van Lente et al., 2003).

However there has been no systematic evaluation of (institutional) intermediary roles to address barriers to (low-carbon) innovation (Howells, 2006; Kivimaa, 2014; Yusuf, 2008) and correspondingly mobilise private finance beyond public research grants. This is what our paper seeks to address by analysing the following research question: *How do institutional innovation intermediaries address the complex set of barriers surrounding (low-carbon) innovation from R&D to commercialisation?*

3.3 Methods and data

To develop an empirically-based perspective in the context of the above reviewed literature, an exploratory, inductive methodology consisting of a multiple case study design (six cases of project managing organisations fulfilling the intermediary role) is applied (Eisenhardt, 1989; Yin, 2009). Within this frame, we chose analytic induction since we base our empirical work on an initial theoretical perspective also referred to as 'abduction' (Mantere, 2008; Patton, 2002). Our iterative process of literature review and empirical data analysis leads to new insights that were not expected previously (Patton, 2002).

The initial theoretical understanding (see chapters 3.2.1 and 3.2.2) revolved around barriers to low-carbon innovation along the innovation cycle and possible policy responses. During the research process we investigated the role, functions and instruments of innovation intermediaries to address these barriers by conducting interviews and a workshop with central intermediary actors in the German innovation system.

3.3.1 Research context

Germany is a particularly interesting research setting due to its strong focus on innovation-led growth, comprehensive environmental regulation and particular financial system. This has consequences for our research design that focuses on financial mobilisation functions of innovation intermediaries in the policy context. Germany plays a leading role in conducting systematic transition towards sustainable energy systems using low-carbon innovations. These conditions necessitate stronger attention. First the industry structure (i.e. large firms, 'Mittelstand') is orientated towards leading-edge technologies. Second, the conservative bank-based system focuses on investment banking and project finance, thus lacking an institutionalised finance system as in the UK or US. Third the strong mission driven government that sets up proactive public policies to overcome existing lock-ins and path dependency.

3.3.2 Sample

The role of intermediaries in addressing barriers to low-carbon innovation and their functions to mobilise private finance in the German innovation system necessitates explanation. Policy making is carried out by the Federal government and the 16 'Länder' governments. R&D activities are conducted by a range of SMEs and larger companies as well as a range of higher education institutions, academies and research organisations (MPG, FhG, HGF, and WGL[10]). Intermediaries between these three parties include the German research foundation, project managing organisations as well as associations and chambers among others. Following the recommended approach to select case studies for analytic induction, we applied theoretical sampling (Eisenhardt, 1989; Yin, 2009). Rather than designing a statistically representative sample, our goal was to select cases that are valuable to investigate (Siggelkow, 2007). Hence we focus on project managing organisations as they occupy a critical intermediary position (see Figure 3-1). These public or private corporations gain their mandates from ministries in a competitive process. With regard to competencies and headcount they surpass their ministerial counterparts and thus play a critical role in bringing public and private actors together. With our sample, we cover six project managing organisations that manage most of Germanys' cooperative R&D projects. They exhibit no clean technology related characteristics. However, to study their role

---

10 MPG: Max-Planck Society; FhG: Fraunhofer Society; HGF: Helmholtz Association; WGL: Scientific Community Gottfried Wilhelm Leibniz.

in the specific context of clean technologies, we focus on 20 government-supported clean technology R&D partnerships at different stages in the innovation cycle (Table 3-3). These projects appertain to the German framework 'Research for Sustainable Development' that aims at fostering clean technology innovation with a technical focus on energy production and efficiency, mobility and materials, amongst others. They also provided deep insights into high-tech government-supported cooperative R&D projects in general. Figure 3-1 depicts an overview of the research setting.

Figure 3-1: Position of the intermediaries and unit of analysis (research setting)

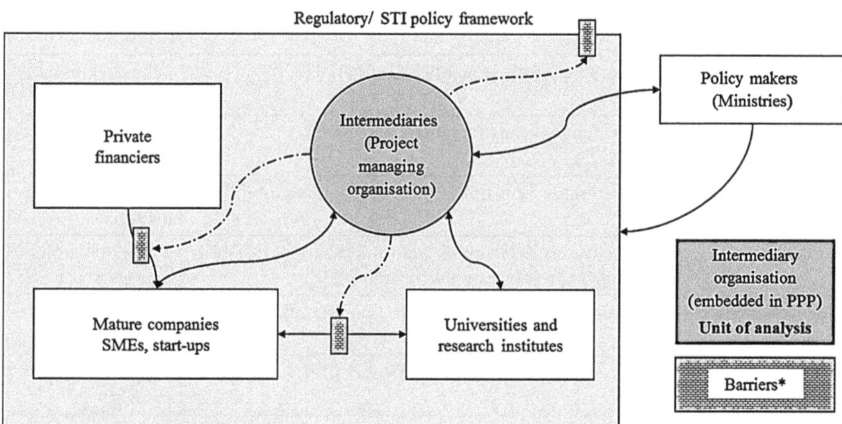

—·— Relationships under investigation

*Technological, regulatory/political, cooperative and financial

For our exploratory study, we contacted leading scientists (individuals) from all project-managing organisations executing the above mentioned 20 R&D partnerships. The surveyed project managers possess broad knowledge about ongoing research process and are aware of the regulatory environment. They are also able to establish links between financiers and supported organisations, as they manage relationships consisting of flows of information and finance within R&D projects. Furthermore, as Figure 3-1 shows, they are not only directly in contact with actors in the innovation system, they are also able to indirectly address barriers to (low-carbon) innovation by capitalising on their position. An overview about the organisations and individual interview participants can be found in Table 3-3.

Table 3-3: R&D partnerships and interviewees

R&D Partnership	Project managing organisation(s)	Interviewees (individuals)
E-Mobility	VDI/VDE Innovation + Technik GmbH	2 Research project managers
Bio-economy	Forschungszentrum Jülich GmbH (PTJ)	Head of research
Biofuels	Forschungszentrum Jülich GmbH (PTJ)	1 Research project manager
Fuel cells	Forschungszentrum Jülich GmbH (PTJ) NOW GmbH	2 Research project managers
Carbon capture and storage	Forschungszentrum Jülich GmbH (PTJ)	2 Research project managers
Carbon nano tubes	Forschungszentrum Jülich GmbH (PTJ)	1 Research project manager
CO_2 Sequestration	Forschungszentrum Jülich GmbH (PTJ)	1 Research project manager
Chemical usage of CO_2	Deutsches Zentrum für Luft- und Raumfahrt (DLR)	2 Research project managers
E-Energy	Deutsches Zentrum für Luft- und Raumfahrt (DLR) Forschungszentrum Jülich GmbH (PTJ)	2 Research project managers
Geo-information	Deutsches Zentrum für Luft- und Raumfahrt (DLR)	2 Research project managers
Geothermal energy	Forschungszentrum Jülich GmbH (PTJ)	1 Research project manager
'Green Carbody Technologies'	Projektträger Forschungszentrum Karlsruhe (PTKA)	1 Research project manager
Advanced materials	Forschungszentrum Jülich GmbH (PTJ)	1 Research project manager
Light emitting diodes (LED)	VDI Technologiezentrum GmbH Forschungszentrum Jülich GmbH (PTJ)	2 Research project managers
Lithium-ion battery	Forschungszentrum Jülich GmbH (PTJ)	1 Research project manager
Organic light emitting-diodes (OLED)	Forschungszentrum Jülich GmbH (PTJ) VDI Technologiezentrum GmbH	2 Research project managers
Organic photovotaics	VDI Technologiezentrum GmbH Forschungszentrum Jülich GmbH (PTJ)	2 Research project managers
Photovoltaics	Forschungszentrum Jülich GmbH (PTJ) VDI Technologiezentrum GmbH	2 Research project managers
'Next Generation Solar Energy'	Forschungszentrum Jülich GmbH (PTJ)	1 Research project manager
Intelligent mobility	TUV Rheinland	1 Research project manager

3.3.3 Data collection

In case study methodology, data collection means reconstructing and gaining understanding at the same time (Flyvbjerg, 2006; Yin, 2009). The overall investigation lasted from mid-July to mid-October 2012. We adopted a procedure whereby we embedded our interviews and a workshop in extensive desktop research on archival documents from multiple sources. These three elements form our cases.

First we reviewed industry reports and reports on climate and innovation policy and we specifically analysed the framework of 'Research for Sustainable Development' for recent and on-going R&D partnerships. These documents (e.g. roadmaps technology documentation, ministerial, participant descriptions) provided basic information about actors, technologies and the innovation process and helped us develop an understanding of how R&D partnerships are structured. For each R&D partnership, a detailed context study with indicative financing needs was developed. The material gathered was also grouped under separate topical headings to enable comparison and cross-linkages. The key characteristics of the R&D partnerships have been very heterogeneous. We distinguished these partnerships in a continuum from early stage (commercialisation) to later stage (diffusion) as they typically require different financial instruments and face different barriers (see Table 3-1). In addition, the volume (2,5 Mio. € – 600 Mio. €) and the ministries responsible for the partnerships as well as the participating actors (SMEs, MNEs, start-ups, universities, research organisations) differed widely.

Second, the descriptions were used in interviews with the experts that took place after a preliminary synthesis and analysis of the written material. The actual interview was important to acquire an in-depth understanding of the peculiarities of specific cases. The number of interviewees per unit varied between 1 and 2 depending on the organisational setup of the R&D partnership, totalling 25 interviews. We conducted the interviews in sequential order to enable the transfer of insights from each case to an improved interview guide for subsequent cases. Each interview took between one and two hours and was conducted face-to-face or via telephone, with one or two researchers present. The interviews were tape recorded with the interviewees' approval and transcribed verbatim. All interviews followed a semi-structured outline, with a set of guiding questions repeated at each interview (see chapter 8.1.2 for the interview guide). The idea was to follow a 'story-telling' approach, to let the interviewees describe their views on the phenomenon and their role as freely as possible, allowing them to interpret the questions and pursue any themes they regarded as central.

Third, the aggregated results were discussed, validated and extended during a workshop which consisted of 15 study participants and 5 R&D project managers in similar positions. The workshop has equally been recorded and transcribed.

3.3.4 Data analysis

To analyse the gathered empirical material we followed the abductive circle of inference. Thus, we deepen our understanding of the meaning of the material in circular movement where the details of a certain text are contrasted with emerging, more generalised theoretical thoughts which involves 'a constant movement back and forth between theory and empirical data' (Gadamer, 1993; Mantere, 2008; Wodak, 2004, p. 200). Starting with our initial theoretical understanding of the barriers to low-carbon innovation along the innovation cycle we have gone through three empirical steps.

First, the extensive desktop research revealed a systemic contextual perspective on the R&D partnerships including the main actors and peculiarities of each field, depending on their stage in the innovation cycle and corresponding technology.

Second, the interview texts were analysed through a hermeneutical research approach (Gadamer, 1993). We applied a combination of narrative and content analysis since we covered most of the aspects asking broad questions, leaving as much room as possible for the interviewee to recount the characteristics of the partnerships, their role in addressing potential barriers (Polkinghorne, 1988) and embedding their views in the context analysis based on the documents (Boje, 2001; Mantere, 2008). Two researchers systematically analyse the data for corresponding patterns and independently mark quotes in the interview protocols, which is referred to as actual process of analytic induction (Eisenhardt, 1989; Yin, 2009).

To infer from the data, we employed axial open coding with emerging categories (Dougherty, 2002) which is guided by an overarching (theoretical) understanding (Patton, 2002). The coding scheme was developed top-down and bottom-up to remain amenable to the emergence of new categories which could still use the axial coding procedure to identify relationships between the main constructs. Following this procedure the first order constructs emerged. The corresponding coding procedure resulted in 1375 quotes from 322 pages (see Table 3-4). We used the software MaxQDA 11 for the text analysis to manage the data.

The relational analysis revealed patterns of codes that occurred repeatedly with each other, and the results were structured around second order constructs (dimensions) to reduce complexity. The dimensions represent factors that influence the perception of barriers by intermediaries and their role in addressing these. The dimension comprise political / regulatory environment, constellation of actors, underlying technology, financing innovation, forms of interaction, intermediaries roles and intermediaries instruments (see Table 3-4). Based on these dimension we carried out a cross-case-analysis on the individual level. The project managing organisation as the unit of analysis did not turn out to be a critical influencing factor (dimension).

Third, the final categories of our analysis have been validated throughout the workshop held (see Table 3-4). The workshop participants focused on the definition of the financing problem for (low-carbon) innovations as well as the contextual dependency of the resulting financing problems which left the question of adequate tangible instruments to address the abstract gap. This workshop setup permitted us to reflect on the role of the intermediaries, to discuss the barriers to low-carbon and to validate the results across industries, which is unique to an intermediary study.

Finally, based on the dimensions above, we wrote micro-narratives integrating document analysis and interview texts, resulting in a holistic perspective on the 20 R&D partnerships which simplified interpretation of the interviewees' opinions against the contextual background made up by the bulk of documentation.

Table 3-4: Qualitative interview coding scheme and category emergence

Freq.	First-order codes	Second order codes (dimensions)	Final categories
44	Regulatory / political barriers	*Political / regulatory environment*	***Sources for barriers to (low-carbon) innovation on the public and private side*** (Foxon & Pearson, 2008; Klein Woolthuis et al., 2005)
40	Policy instruments		
49	Policy environment (STI and regulation)		
47	Interfaces (between major actors and phases in the innovation cycle)		
113	Characteristics of the R&D partnership (i.e. Strategy, roadmaps, goals)	*Constellation of actors*	
37	Technological barriers	*Underlying technology*	
32	Commercialisation (of the respective technology)		
25	Properties of the technology		
77	Subsidies (given to research institutes and firms)	*Financing innovation*	
50	Financial barriers		
32	Entrepreneurship (activities of universities and research institutes)		
25	Own involvement (how intermediaries execute related finance tasks)		
50	Public-private cooperation	*Forms of interaction*	
19	Private-private cooperation		
32	Intermediaries competencies and mandates (general)	*Intermediaries roles*	***Role of intermediaries in addressing barriers*** (Howells, 2006; Kivimaa, 2014; Klerkx & Leeuwis, 2009)
11	Administration		
38	Manager		
32	Consultant		
29	Bridge builder		
41	Limitations		
13	Demonstration	*Intermediaries instruments*	
21	Socio-economic tools		
22	Advisory opinion		
29	Other tools		

To ensure the robustness of our analysis a number of measures have been taken (Creswell & Miller, 2000; Moran-Ellis et al., 2006; Patton, 2002): First, to ensure construct validity (establishing correct measures) several data sources have been used (archival documents, notes, interview transcripts and workshop transcripts) (Patton, 2002). After analysing all the cases, we asked the project managers to review the key results from their individual technology fields at the workshop held to detect and avoid potential misunderstandings ('member checking') (Creswell & Miller, 2000). Second, to ensure internal validity (avoiding alternative explanations) we compiled a representative sample (including all major project managing organisations in Germany) which includes typical cases across of a large set of technologies covered by the R&D partnerships This approach allows us to add to existing theory (Moran-Ellis et al., 2006). Third reliability was ensured by conducting the research with two scholars, by developing and refining the semi-structured interview guide for both interviews and workshop, by recording and transcribing the evidence, by developing a case database and by finally writing micro-narratives.

3.4 Results

Based on the hermeneutical and relational analysis and the emergent categories we identified financial barriers to (low-carbon) innovation that are linked to other forms of barriers which are typically addressed by innovation intermediaries such as regulatory, cooperative and knowledge or information barriers. We then explore the role of institutional intermediaries in addressing these barriers, focusing on their financial mobilisation functions. Finally we argue that based on their role and the use of these functions intermediaries exercise an influence on the innovation process of selected technologies. Representative, numbered quotes for the following argumentation can be found in Table 8-1 (Appendix).

3.4.1 An intermediaries' perspective on the relation between financial barriers and other barriers to (low-carbon) innovation

On the one hand, practically every field of innovation sets out goals towards commercialisation and application in Germany, while some build on sophisticated roadmaps as part of their strategy (e.g. electric mobility or organic electronics). However only a few of them specified instruments to tackle the transition from basic research over applied R&D towards commercialisation. On the other hand, the interviewees were returning to the perceived barriers to low-carbon innovation along the innovation cycle as a major theme. From an intermediary position this perspective provides valuable insights. The analysis of the interview texts however revealed that financial barriers often pertain to other aspects such as the political environment and corresponding barriers, technological barriers, commercialisation, information asymmetries, intermediaries competences and limitations and public-private cooperation depending on the phase in the innovation cycle.

In the phase of basic research the R&D project managers highlighted technological barriers such as complexity and administrative barriers such criteria for the selection of projects (B1, B2). These translated directly into constrains regarding funding (i.e. grants), as well as potential sources of finance for further development (B3, B4). Upon entering the applied R&D phase, the experts stressed the missing orientation towards commercialisation, both on the private side such as slow adoption of technologies, missing cooperation along the value chain, low corporate R&D and on the public side i.e. limited ability to make the technological development process transparent, missing links between STI and industrial policy as well as public-private interaction (commercial viability or business models for technologies) (B6, B8, B9). These barriers directly relate to missing private risk capital (i.e. business angels or VC) and uncertainty about further public funding for the technologies under development (B5, B7).

Throughout the demonstration and pre-commercial phases the R&D project managers referred to severe financial barriers on the private side e.g. capital intensity and missing collateral, bankability, insurability, competence problems and too short time horizons for financiers (B10, B12). On the public side, administrative barriers such as missing interfaces between phases in the innovation cycle and government bodies and a limited availability to address this gap by the intermediaries themselves were highlighted (B11). Finally institutional barriers such as infrastructure occurred (B13).

Finally, for the niche-market, supported commercial and the fully commercial phase, the interviewees perceived a high path dependency on the private side regarding technologies and business models (B17) but also high regulatory and political uncertainty and inconsistent support for different technologies (i.e. picking the winner problem) on the public side (B14, B15, B16) Public-private barriers arise from too early commercialisation effort as a result of technology push measures and responding private actors as well as infrastructure problems (B18). In addition the compatibility between private finance and public support initiatives is lacking (B19).

Drawing from the perspective of institutional intermediaries at the intersection between public and private actors the results indicate that technological, co-operative and political barriers along the innovation cycle directly translate into financial barriers. This relationship takes different forms along the innovation cycle. In the early stages of the innovation cycle technological and administrative barriers such as opaque technology development process, commercial viability limit the potential matching with funding sources. A missing orientation towards commercialisation including a focus on costs and potential business models and missing interfaces between ministries and stages in the innovation cycle combined with limited capabilities of financiers translates into a severe gap for private risk capital during the demonstration and niche-market phases. During the later stages in the innovation cycle (supported commercial and fully commercial) regulatory and policy uncertainty as well as inconsistent support mechanisms represent the most persistent barriers, which translate into large investments (projects and infrastructure) been delayed or withdrawn.

3.4.2 Role of institutional intermediaries in addressing these barriers

Having lined out the perceived technological, cooperative, regulatory or political and corresponding financial barriers we analysed the role of institutional intermediaries to address these barriers and correspondingly mobilise private finance which is inhibited by the barriers.

Competencies and mandates regarding commercialisation
Although the intermediaries institutionally fulfil the same position, we found a variety of roles, competencies and mandates that impact their behaviour within their respective contexts. The surveyed project managers see themselves as experts – a prerequisite for fulfilling managerial task of the R&D partnerships (C1). Secondly, they act as bridge-builders between critical actors in the innovation system – within the supported R&D projects and the corresponding participants (C2, C3, and C5) that addresses cooperation barriers in the early stages. However, only few explicitly regard themselves as organisations designed to bridge the gap between R&D and commercialisation respectively address underlying financial, regulatory and technological barriers (C4).

The R&D project managers allocate resources between projects canalising public funds, selecting participants and documenting and controlling the process as well as the usage of the generated knowledge (products, processes or patents). Some of them do not see a relevant role within that process (C6), whereas others have a holistic understanding of their managerial capabilities (C7). Most surveyed project managers have a broader commercial perspective which is supported by the analysis of archival documents for each field. However, only a few directly address barriers such as capital intensity and scalability on the private and the missing provision of infrastructure on the public side.

Example: 'LED Lighting R&D and Lead-market Initiative'

The LED has been government supported since the potential for general lighting was found at the end of the 1990s. The project management organisation and individual project managers financed *basic and applied research* pushing the technology from small scale to large scale application along the technological life cycle with a focus on interfaces between ministries and stages. The aim was to provide the German LED manufacturers with possibility to cope with the transition in the lighting industry (e.g. Solid State Lighting – SSL). Thus mostly *mature companies, research institutes and universities* benefited from this program. At the verge of commercialisation in 2009, the ministry together with the project managing organisation launched the LED lead market initiative which aimed at accelerating the commercialisation and diffusion. This initiative was an *institutionalised public-private cooperation* coordinated by the project managing organisation. Here the project managers acted as *bridge builders and innovation managers with a clear commercialisation perspective*. Main topics include the financial and organisational barriers to innovation such as risk, and uncertainty regarding business models based on LED. These barriers have been evaluated against the current institutional and regulatory environment, thus feedback to design conducive regulations was provided by the project managing organisation. Throughout the lead market program, *demonstration projects, application programs and well as public-private instruments* (e.g. standard energy service contracts for lighting application) have been developed. Further results include *guidelines for LED modernisation and certificates* to ensure quality.

Supporting policy makers to influence the financing environment
Due to their contractual relationship with government ministries and the resulting innovation agent characteristic of the institutional intermediaries, a major role revolves around institutionalised dialogue with political actors. In this respect, they act as consultants influencing the design of R&D programs (P1, P2). Resulting from their critical position between policy-makers and industry, the surveyed managers also engaged deeply with the review and extension of existing regulations (P3, P4). Most study participants established the link between policy environment and their technology field, a few with consequences on advanced eco-systems, especially the finance environment (P5). However, some still regard their influence on accompanying regulatory environment as limited (P6), especially in the niche-market and supported commercial phases.

Within our narrative analysis, the interfaces between different stages of the innovation cycle that leave structural holes such as the 'valley of death', the responsible actors for the situation and the appropriate instruments to correct these failures returned as a major themes. This includes the responsibilities of different ministries along the innovation cycle as well as coordination of support mechanisms (P7, P8). Most of the intermediaries address the structural holes in the innovation process, notably the transition from basic R&D to applied R&D and the 'valley of death' at the edge towards commercialisation (P9). Still, several R&D managers did not see the transition towards commercialisation as their responsibility or even considered their neutral position within the innovation system endangered (P10).

Financial instruments and cooperation with financiers
The R&D project managers possess a set of instruments to support the innovation process respectively to address the above mentioned barriers. They include varying subsidies schemes, socio-economic research and start-up support schemes, as well as instruments targeting the later stages of the innovation process, e.g. commercialisation which requires contact with public or commercial funds. These contacts might take various forms and are seen as a new competence for R&D project managers (F1, F2). Additionally supporting tools such as roadmaps were highlighted. These tools make the technology development process (including complementary assets such as infrastructure) understandable to third parties and help in coordinating actors (F3).

Others argue in favour of integrating VC as an instrument for effective support of SMEs and start-ups, to signal quality towards potential financiers (F4) as external capital is often missing. However information used as signal for potential financiers or other private actors might be confidential or unavailable in aggregated form (F5). Supplying these materials could mean a conflict of interests for the intermediaries (F6). Still, R&D project managers could provide complementary research to reduce the risks of private financiers (F7). In sum, two possibilities to address financial barriers emerged. The intermediaries could either directly use their instruments to address financial barriers, especially relevant in the presence of SMEs or they could shape the policy environment address regulatory risks and uncertainty in the later stages in order to provide incentives for larger companies to invest into innovative capacities.

During the workshop held, the participants highlighted their systemic perspective and their financial mobilisation functions to address specific structural gaps and financing needs. Two solutions have been revealed. First, the integration of market perspective in the design of R&D programs was considered, that would permit a smooth transition between R&D and commercialisation. Second, they highlighted their bridge-building and gatekeeping functions later in the innovation process to provide interfaces between the market participants.

Example: 'Smart-grids innovation program (E-Energy)'

The smart grids innovation program aimed at a demonstration of feasibility using existing and novel clean technologies in intelligent model regions. Thus the initiative focused explicitly on the *outset between applied R&D and commercialisation*. It specifically addressed *SME and start-ups* which have been particularly active in this cleantech subsector. The project managing organisation took an *innovation management approach*, focusing on problems that SME highlighted as barriers for an accelerated commercialisation: Missing business models and the limited access to VC for a quick scale up. They consequently applied instruments such as *prizes for founders and a holistic model region approach*. The combination of *VC and R&D grants* as financial instruments has not yet been implemented, although the responsible project manager framed it as a possibility to address financial barriers. *Standardisation* as a technological barrier was further translated back to the corresponding ministries to be addressed on a regulatory level.

3.5 Discussion and conclusions

The research question guiding our enquiry was: How do institutional innovation intermediaries to address the complex set of barriers surrounding (low-carbon) innovation from R&D to commercialisation? In this chapter we will reflect upon this research question and mirror our findings to previous insights from the literature on innovation finance and innovation intermediation, to show the specific role of intermediaries in mobilising finance and how this relates to the addressing of barriers to innovation.

3.5.1 Addressing financial barriers requires a holistic perspective

Due to high public-private uncertainty, R&D complexity and learning in the early stages low-carbon innovations need increased support, coordination of activities and the development of complementary assets (e.g. infrastructure, standards, etc.) (Haley & Schuler, 2011; Kenney & Hargadon, 2012; Mathews et al., 2010). Thus, especially in the case of low-carbon innovation more systemic efforts are needed to escape lock-in effects and path-dependency and to accelerate the commercialisation and diffusion of clean technologies by balancing regulation, innovation and complementary financial mechanisms. According to our results, institutional innovation intermediaries possess the instruments that address the underlying financial barriers along the innovation cycle, such as capital intensity, scalability, infrastructure, lock-ins, regulatory risk and policy coordination failures.

Our findings complement earlier work that has been looking at single public and private instruments to finance innovation (Auerswald & Branscomb, 2003), notably research programs and grants (SBIR / ATP) (Chang et al., 2002; Link &

Scott, 2010) and VC or business angels (Kenney & Hargadon, 2012; Leete et al., 2013; Nightingale et al., 2009). Our research confirms the utility of individual funding instruments; however we highlight restrictions of these instruments with regard to their coordination and embeddedness within the overall innovation system and process, in order to thicken up the thin financial markets for (low-carbon) innovation. This requires a focus on the interfaces between the innovation process stages and its connection with the policy environment. So coordination of the variety of policy instruments aimed at financing innovation by an innovation intermediary addresses the lack of specific focus on financial mobilisation intermediation in many innovation systems, which results in many technologies failing to reach the market (Bergek et al., 2008; Jacobsson & Karltorp, 2013; Mathews et al., 2010). Our findings hence provide empirical support for the argument put forward by Wonglimpiyarat (2011), who found that the 'financial innovation system' underlying the national and technological innovation systems is a significant driver of innovative activity. We show that financial barriers interact with technological as well regulatory barriers along the innovation cycle.

Hence financing (low-carbon) innovation requires a holistic perspective on the innovation cycle as each phase requires different forms of financing and support to address the underlying financial and non-financial barriers along this cycle. In the next chapter we discuss the key strategies by institutional intermediaries that emerged from our results.

3.5.2 Exploring the financial mobilisation functions of (institutional) innovation intermediaries

Based on our analysis of intermediary roles and instruments we derived a set of functions that permit institutional intermediaries to influence the finance environment for (clean) technologies and consequently accelerate the commercialisation and diffusion process. 'Financial mobilisation functions' comprise not only strictly financial instruments but also instruments that indirectly influence the finance environment for (clean) technologies. Table 3-5 provides an overview and we will highlight the four main components below.

Firstly 'classical' innovation intermediation functions indirectly impact the financing environment, as intermediaries are able to accelerate the commercialisation process by establishing and managing strategic research partnerships and by using supportive instruments such as roadmaps, strategic public procurement or production support measures. Previous literature on innovation intermediation has highlighted the direct impact of these instruments upon the innovation process (Edquist & Zabala-Iturriagagoitia, 2012; Kivimaa, 2014; Klerkx & Leeuwis, 2009; Link & Scott, 2010). Based on these results, we hence argue that these instruments determine the attractiveness for private financiers to invest into companies and complementary assets.

Secondly, the results highlight support functions for STI policy and regulation surrounding the technologies under development. In addition to the beneficial effects of adequate STI policy and regulation for innovation and commercialisation (Kivimaa, 2014; Klerkx & Leeuwis, 2009; van Lente et al., 2003), this indicates that support for favourable STI policy mechanisms and regulation directly determine the ability of private investors to invest into young, small and more mature companies. These investments are especially relevant in the context of low-carbon innovation as these technologies exhibit a strong regulatory dependency and asset heaviness.

The third component of 'financial mobilisation functions' revolves around the cooperation with private financiers to improve their competences regarding innovative (clean) technologies and thus strengthen their ability to evaluate future market opportunities. Reduction of information asymmetries has been highlighted in the context of markets for technologies (between producers and users) (Hoppe & Ozdenoren, 2005); through our analysis we found that it is especially relevant in the context of financing (low-carbon) innovation since the information asymmetries between financiers and innovators are greater.

Fourth, direct financial instruments represent probably the most obvious part of financial mobilisation functions to alleviate financial constraints for innovating firms and research institutes. These include subsidies, grants, tax credits and support for demonstration projects (Howells, 2006; Yusuf, 2008). Our analysis adds to this earlier work by indicating that it is critical to supply the adequate amount of finance to allow a seamless transition between the phases of the innovation cycle. This includes a combination of public (e.g. grants) and private (e.g. VC) instruments in later stages to leverage the publicly invested money.

Table 3-5: Financial mobilisation functions and corresponding instruments

Financial mobilisation function	Instruments
Competencies and mandates of innovation intermediaries	Knowledge and technology brokerage (Roadmaps, research program design etc.) R&D projects covering multiple elements of the value chain Strategic research partnerships (e.g. SBIR, ATP, DARPA, …) Public procurement or production support measures
Policy support: Regulation and STI policy	Support for STI policy Support for regulation Complementary assets (infrastructure etc.)
Public-private cooperation (e.g. with investors)	Mobilise private finance (Business angels, VC, Private equity, banks, mezzanine) Improve positive expectations of future market opportunities Support for competence building in VC/PE, Banks etc.
Financial instruments	Subsidies, grants Tax credits Demonstration projects Combination of support and VC/PE

3.5.3 The synergic role of financial mobilisation intermediation with other intermediation functions

As discussed in chapter 3.5.2, we argue that in order to address the interwoven barriers highlighted above, intermediaries need to apply different roles and corresponding instruments for each stage of the innovation cycle, hence we call for a context-dependent intermediation. We highlight the fact that financial mobilisation functions represent critical instruments to support and accelerate the innovation process for clean technologies by intermediating between public policy makers, private financiers and innovators. From our results we derived a model which is depicted in Figure 3-2.

Figure 3-2: Model for intermediaries to address barriers to (low-carbon) innovation

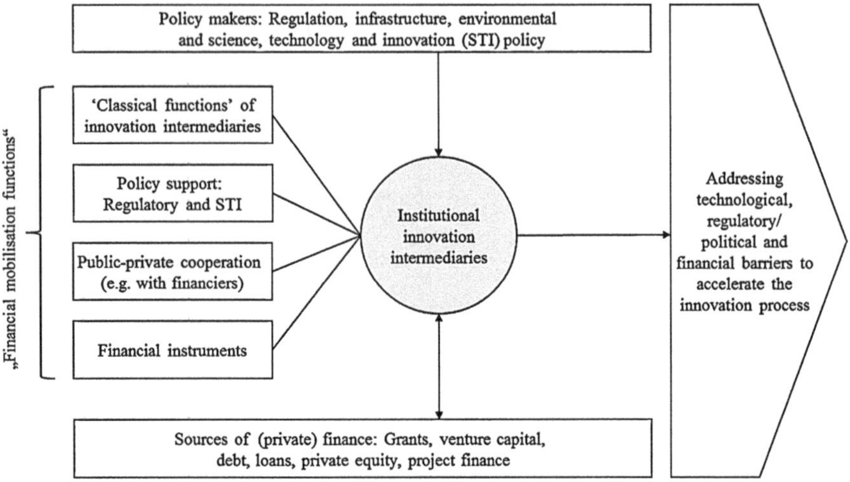

These different roles and functions are synergic, as becomes clear from the results in different ways. First, and this could be called a 'first-order effect', they capitalise on their strength of public-private cooperation (i.e. innovation intermediation – acting as expert, bridge builder, innovation manager) focussing on bottlenecks along the innovation process to bridge the 'valley of death' and other structural holes which goes beyond funding basic or applied R&D and demonstration. Howells (2006) described this as generic role throughout the commercialisation phase. Institutional intermediaries coordinate different sources of capital in the early stages (e.g. research grants, PPP, private sources of capital) that represent the direct financial instruments component of our concept. The focus on bottlenecks additionally addresses barriers such as policy coordination and reflexivity failures, highlighted by Weber & Rohracher (2012).

Second, as a 'second-order effect' they accelerate the innovation process by supporting the design of the policy environment that is conducive to the innovation process spanning from complementary assets towards regulation to indirectly change the finance environment (Klerkx & Leeuwis, 2009). This supports work that argues that as a whole financial mobilisation functions could act as a lever for publicly invested money and permits them to thicken up thin finance markets for (low-carbon) innovations (Kleer, 2010; Nightingale et al., 2009). Our results reveal that a technology specific assessment of policies is necessary to facilitate financing (cleantech) innovation. The development of complementary assets (e.g. infrastructure, standards) or research to reduce risks enables private financiers to more seamlessly invest into the commercialisation phase of (clean) technologies and thereby overcome corresponding barriers.

Thirdly, there is a synergy between public and private actors in the sphere of innovation finance, as the intermediary bodies take the form of government funded institutional agents which act in a neutral way between market and non-market actors. This form of intermediary, complements informed investors (which to some extent may also be considered intermediaries) such as VCs (Da Rin et al., 2006; Leete et al., 2013; Repullo & Suarez, 2000) as they possess the capability of reducing uncertainty and risk. Financiers (such as VCs and banks) are not able to grasp the complexity of the whole innovation processes, systemic innovation, or only perform certain types of investments such as complementary assets (e.g. infrastructure), especially for clean technologies. Hence, here the link between VC and institutional intermediaries takes the form of what Stewart and Hyssalo (2008) have dubbed an 'ecology of intermediaries'.

3.6 Conclusion and implications for research and policy

Based on the findings of this study, we show that institutional intermediaries with their 'financial mobilisation functions' accelerate the commercialisation and diffusion of (low-carbon) innovation. These functions may catalyse other innovation policy instruments, and vice versa. The coupling of policy instruments for supporting innovation and private finance via institutional innovation intermediaries addresses the missing financial 'financial mobilisation intermediation' (Bergek et al., 2008; Jacobsson & Karltorp, 2013; Mathews et al., 2010). We therefore contribute an intersection between the financing innovation literature and the innovation intermediaries literature (Howells, 2006; Klerkx & Leeuwis, 2009; Wonglimpiyarat, 2011).

From our study, two important policy recommendations can be derived. First of all policy makers should consider (financial) barriers and related technological, cooperative and regulatory or political barriers as intertwined. By ignoring the linkages between financial and other barriers, i.e. considering innovation and finance separately policy makers lose the potential to support critical industries such as cleantech in an efficient and effective way. Second, institutional intermediaries with specific financial mobilisation functions represent a crucial instrument for policy makers to operationalise their policy measures and enhance eco-innovation. Policy makers should therefore extend the mandate of institutional intermediaries and capitalise on their position in the innovation system to help designing an optimal mix of policy instruments, to optimally coordinate public support for the development of technologies and to facilitate exchange between financiers and innovators at an early stage.

Since intermediation is a context dependent phenomenon, we acknowledge limitations to our study. The study's qualitative nature prevents us from generalising results. Still, the surveyed partnerships represent typical cases for high technology R&D. Thus the model based on our findings could be transferred to other forms of organisations or government agencies in other contexts, as the functions remain universal. Organisations that have similar responsibilities and face similar barriers exist in other countries. They include the Department of Energy National Laboratories in the US, the Finnish Funding Agency for Technology and Innovation (TEKES) or the Department of Trade and Industry (DTI) in the UK. However, depending on degree of influence from institutional conditions i.e. structural R&D support in the national innovation system, the transferability of our findings might be limited. To gain deeper understanding of the interaction between institutional intermediaries, innovators and financiers, additional studies need to be conducted in other highly uncertain fields of technology.

Acknowledgements
The authors are grateful for the time and support of the interviewees and workshop participants. Valuable comments on earlier versions of the paper have been provided by Florian Täube and Christian Friebe (EBS Business School), Paul Nightingale (SPRU – University of Sussex) and Paula Kivimaa (Finnish Environment Institute SYKE) as well as three anonymous conference reviewers. The opportunity to present and discuss earlier versions of the paper at oikos Young Scholars Entrepreneurship Academy 2012, DRUID Summer Conference 2013 and Academy of Management Conference 2013 helped us tremendously in further refining our argument. The research team would like to thank the Federal Ministry of Education and Research (BMBF), Germany, for their financial support as part of the research project "Climate Change, Financial Markets and Innovation (CFI)".

4 Exploring the role of ESCOs to overcome barriers for innovative energy demand reduction technologies – the case of public LED street lighting in German municipalities

Friedemann Polzin, Paschen von Flotow & Colin Nolden

Abstract: Street lighting in Germany is primarily the responsibility of municipalities. The novelty of LEDs and their challenging diffusion can exceed existing municipal lighting governance capacities. Drawing from the energy services and product-service-system literature, this paper analyses the role of ESCOs in addressing some of the technological, economic, institutional and competency barriers currently hindering more rapid technological diffusion using transaction cost economics. The analysis provides recommendations regarding the governance of municipal street lighting and how ESCO solutions may help challenge incumbent structures.

4.1 Introduction

Energy efficiency is the cheapest and most effective way of addressing issues such as climate change and energy security (EC, 2014; IEA, 2014a, 2014b; IPCC, 2014). However, investments in associated technologies remain below the optimal level as the markets for energy efficiency are failing (IEA, 2013b). Consequently, the rate of adoption and diffusion is low as structures and incentives for application are missing (Jaffe et al., 2005; Schleich, 2009; Sorrell, O'Malley, Schleich, & Scott, 2004). In this paper we analyse the case of public sector application of LED (light-emitting diodes) street lighting in Germany as an example of a market where some of the barriers are being addressed through ESCO solutions.

The lighting industry has recently undergone major shifts from traditional lamps towards LED with significant savings in terms of finance and energy (IEA, 2013a). As a result, application of this technology is proving to be challenging for both manufacturers and customers (Bergek & Onufrey, 2013; Sanderson & Simons, 2014; Smink et al., 2013), despite forecasts of LED market shares increasing from less than 10% to 70% by 2020 (McKinsey, 2012). Public procurement is considered a significant driver in the innovation and diffusion process (Edler & Georghiou, 2007; Edquist & Zabala-Iturriagagoitia, 2012) and in this paper we analyse the role of different governance arrangements available to German municipalities for the procurement and management of LED street lighting.

Street lighting in Germany represents a major cost factor, accounting for almost one third of municipal electricity budgets (DStGB, 2010). This provides a strong incentive for municipalities to take the lead in the diffusion of more efficient technologies (Radulovic, Skok, & Kirincic, 2011; Schönberger, 2013). German municipalities, however, are typically short in budgets with total municipal debt amounting to €133.6bn in 2013 (Difu, 2014; DStGB, 2014). Investing in energy efficiency can alleviate financial constraints and help municipalities meet climate change targets (Bennett & Iossa, 2006; dena, 2015; Hartmann et al., 2014; Hendricks, 2014) but overburdening debt often limits capacities to seek and engage in energy efficiency projects, especially those associated with high transaction costs. 'Servitizing' public sector LED street lighting through energy service contracts offered by energy service companies (ESCOs) can help overcome some of these barriers.

Drawing from research on ESCO markets (Bertoldi, Boza-Kiss, Panel, & Labanca, 2014; Marino, Bertoldi, & Rezessy, 2010; Marino, Bertoldi, Rezessy, & Boza-Kiss, 2011) and related concepts such as 'servitization' through product service systems (Maxwell, Sheate, & van der Vorst, 2006; Maxwell & van der Vorst, 2003; Steinberger, van Niel, & Bourg, 2009), this paper applies transaction cost economics (TCE) to the case study of LED street lighting in Germany. It argues that, in the absence appropriate municipal governance structures capable of diffusing LED street lighting, ESCO solutions can accelerate the commercialisation and diffusion of EUEDs. The example of Germany is particularly revealing as municipal independence and its federal state structure have lent themselves to the establishment of diverse governance mechanisms for the provision of public services. Our research questions reads as follows: *What is the role of ESCO solutions in addressing the barriers to commercialisation and diffusion of LED street lighting in Germany?*

Our research combines a longitudinal archival document analysis with 40 semi-structured interviews. Chapter 4.2 reviews the relevant literature and introduces the theoretical framework. The research design including the qualitative research approach and the data is presented in chapter 4.3. Chapter 4.4 displays the results which form the basis for discussion (Chapter 4.5) and conclusion (Chapter 4.6), including policy and management implications.

4.2 Theoretical background

4.2.1 Challenges facing adoption of novel end-use energy demand technologies

One way to address climate change by uncoupling economic growth from the use of finite resources is low-carbon innovation (Foxon et al., 2008; Popp, 2010). Compared to supply side low-carbon technologies, end-use energy demand reduction technologies (EUEDs)[11] have been marginalised, despite their diffusion considered essential to reach climate change targets while reducing costs and fossil-fuel dependency (Cagno & Trianni, 2014; IEA, 2014a, 2014b; Mickwitz et al., 2008; Sovacool, 2009b). Wilson et al. (2012) attribute this to the nature of these technologies, such as their diversity and widespread application, small scale and low visibility.

Barriers to the diffusion process arise from complex interdependent factors that relate to the nature of innovation and environmental externalities. Technological factors comprise uncertainty about the dominant design, quality, increased complexity of innovative technologies and application (Schleich, 2009; Sorrell et al., 2004; van Soest & Bulte, 2001). Institutional barriers such as path-dependent technological application associated with investments into corresponding infrastructure, low acceptance among the local population or unanticipated or reoccurring changes in the policy design and administrative approval also impede the technological diffusion process (Foxon et al., 2008; Iyer et al., 2013; Klein Woolthuis et al., 2005; Wilson et al., 2012). Volatile or artificially low energy prices and incomplete carbon markets represent economic barriers to diffusion by increasing uncertainty and rendering investments in innovative EUEDs unprofitable, which lead to long payback periods as upfront costs are high (Gallagher et al., 2006; Jaffe et al., 2005; Sorrell et al., 2004). To examine and overcome these factors, potential users require enhanced competencies and capacities (Jacobsson & Karltorp, 2013; Klein Woolthuis et al., 2005; Schleich, 2009; Sorrell et al., 2004). As a result, customers often 'wait' for future improvements and fail to harness current savings. This is referred to as 'energy efficiency paradox' (van Soest & Bulte, 2001). The above-mentioned factors represent key barriers for a range of EUEDs, including lighting (Mills & Schleich, 2014), which this paper seeks to address.

---

11 An 'end-use energy demand reduction technology' can be defined as 'the reduction of the absolute consumption of primary energy' (e.g. electricity, fuel) (Herring, 2006; Wilson, Grubler, Gallagher, & Nemet, 2012); see also: http://www.eued.ac.uk/whatiseued.

The diffusion of novel technologies is heavily reliant on demand-side measures, among which public procurement is one key element (Edler & Georghiou, 2007; Guerzoni & Raiteri, 2014; Testa, Annunziata, Iraldo, & Frey, 2014). Public procurement can also contribute to satisfying unsatisfied human needs and solving societal problems, such as climate change, if appropriate innovations are supported (Edquist & Zabala-Iturriagagoitia, 2012). However, this process is subject to a set of barriers perceived by suppliers of (green) innovative goods, such as the lack of interaction with procuring organisations, the use of over-specified tenders, missing environmental criteria, low competencies of the procurers and a poor management of risk during the procurement process (Testa et al., 2014; Uyarra et al., 2014).

Governments, like municipalities, are typically running on a very tight budget, which limits their abilities to procure innovative EUEDs (Edquist & Zabala-Iturriagagoitia, 2012; Hartmann et al., 2014; Testa et al., 2014). In this case, third-party contracting, a specific form of public procurement, can be a solution (Bennett & Iossa, 2006; Hartmann et al., 2014; Helle, 1997; Hypko, Tilebein, & Gleich, 2010a; Roehrich, Lewis, & George, 2014; Toffel, 2002). The literature has highlighted the design of contractual arrangements (*mode of governance*) regarding contract duration and responsibilities as a critical factor to address the complexity in the procurement process (Hartmann et al., 2014; Roehrich & Lewis, 2014). It has been suggested that public agencies need to 'identify the procurement level and the contractual and relational challenges involved' (Hartmann et al., 2014, p. 174) in a public-private partnership (PPP) to increase the economic viability and ensure the appropriate distribution of benefits.

4.2.2 ESCO solutions for the diffusion of end-use energy demand technologies

Third-party contractors such as ESCOs can be efficient suppliers, especially when it comes to energy-related products and services (Helle, 1997; Sorrell, 2007). This paper focuses specifically on energy performance contracts (EPCs), which are referred to as ESCO solutions. The ESCO assumes control over the secondary conversion and control equipment as part of an EPC, which allows the ESCO to identify, deliver and maintain savings using guarantees for certain standards (of lighting service) at a cost typically lower than its customers' current or projected energy bill (Hannon et al., 2013; Marino et al., 2011; Sorrell, 2005, 2007).

A tendency towards outsourcing, rising energy prices, continuing market liberalisation and environmental concerns benefit ESCO markets (Marino et al., 2011; Zhang, Wu, Feng, & Xu, 2014). Before an ESCO assumes control over conversion equipment during an EPC (see Sorrell, 2007 for more details) and contributes to the accelerated uptake of innovative EUEDs, technological, competency and economic or investment barriers need to be addressed. For example, an ESCO may be able to reduce the cost for energy by deploying (novel) EUEDs and management systems but implementation is typically hampered by a lack of awareness, low priority, non-standardised measurement and verification (M&V) of energy savings and general uncertainty regarding energy efficiency investments, legal complexities, volatile energy prices, missing energy cost information, difficult access to finance, business risk and mistrust against ESCOs (Marino et al., 2011; Pätäri & Sinkkonen, 2014).

Despite these barriers, ESCOs co-evolve with a transition towards a more sustainable energy system (Hannon et al., 2013). However, their success in the German street lighting sector also depends on municipal risk-aversion, the extent to which municipals wants to retain strategic control and the resources they have at their disposal (Hannon and Bolton, 2014).

4.2.3 Research gaps identified in the literature

First, with our analysis we complement research on the adoption and diffusion of innovative EUED, focusing on one specific technology, LEDs, with a large potential for application. LED lighting has been analysed at industry level (Sanderson & Simons, 2014; Smink et al., 2013) and household level, focusing specifically on adoption (Mills & Schleich, 2014). The distinct context of public application has been neglected, although the energy costs municipalities and other public actors incur are significant.

Second, we address the calls for research by Pätäri and Sinkkonen (2014), Roehrich et al. (2014) and Hannon and Bolton (2014) into successes and failures of ESCO solutions in order to derive success factors for their application in the context of PPP and public procurement. ESCO solutions have been shown to be suitable for reducing energy costs for some energy services (Helle, 1997; Pätäri & Sinkkonen, 2014; Sorrell, 2007) and to exhibit elements of a transformative power towards an alternative market design of a performance-based economy (Haas et al., 2008; Hannon et al., 2013; Steinberger et al., 2009). The specific potential of ESCO solutions to accelerate the commercialisation and diffusion of *novel* EUEDs has been neglected in the literature (Hypko et al., 2010a; Roehrich et al., 2014; Steinberger et al., 2009).

4.2.4 Analytical Framework: Transaction cost economics

Transaction cost economics (TCE) represent a prominent theoretical perspective to analyse modes of governance and institutional structures (see Figure 4-1) between hierarchies and markets (Delmas, 1999; Selviaridis & Wynstra, 2014; Williamson, 1985). Transaction costs (TCs) depend on how the transaction is organised through governance structures and TCE makes 'several key assumptions about managerial behaviour when determining which governance structure is most efficient for a particular transaction' (Pint & Baldwin, 1997; Toffel, 2002, p. 2).

TCs are incurred within organisations through managing and monitoring personnel, procuring inputs and capital investment, *the costs associated with organising ('governing') the provision of [...] streams and/or services* (Sorrell, 2007, p. 512). When the same streams and/or services are sourced from an external provider, transaction costs are associated with source selection, contract management and performance monitoring, dispute resolution and opportunistic behavior (Pint & Baldwin, 1997; Sorrell, 2007). Transactions are governed through structures, which are located on a spectrum with hierarchical organisations (internal) at one end and spot markets (external) at the other with hybrid mechanisms in between (Pint & Baldwin, 1997; Toffel, 2002) (see Figure 4-1).

Figure 4-1: Spectrum of governance structures

Vertical integration	Relational contracts	Long-term contracts	Simple short-term contracts	Spot markets

Hierarchies   Markets

Adapted from Pint and Baldwin (1997, p. 4).

With increasingly complex customisation of services, more hierarchical governance structures may be appropriate to allow for adjustments to contingencies (such as the emergence of more efficient technologies) over the life of a contract (Pint & Baldwin, 1997). According to TCE, the choice between long-term contracts (EPCs) and hierarchies (in-house provision) depends on the magnitude of associated transaction (TCs) and production costs (PCs). We employ a model originally developed by Williamson (1985) and specified by Toffel (2002), Pint and Baldwin (1997) and Sorrell (2007). Sorrell's model in particular focuses on the economics of energy service contracts, which allows the analysis of ESCO solutions vis-à-vis other modes of governance in Germany's municipal street lighting market, their capacity to diffuse innovative EUEDs and enhance modernisation.

According to TCE, the choice of governance structure also depends on characteristics of transactions as investment in transaction-specific (dedicated) assets may improve the efficiency of some transactions (Pint & Baldwin, 1997). An

ESCO offering EPCs for lighting, for instance, may already have acquired skills and established links with manufacturers and appropriate equipment for both operation and maintenance (O&M) and measurement and verification (M&V). Acquired skills and the buyer-seller relationship can be classified as *human capital specificity* while specialised equipment falls under *technical asset specificity*. *Dedicated resources* is the degree to which institutions support a particular (Pint & Baldwin, 1997; Toffel, 2002).

The assumption is that transactions featuring *high technical asset specificity* and *dedicated resources* are less likely to operate efficiently within market transactions as the party that has not invested, in this case the municipality, may threaten to cancel the contract to expropriate some of the invested value as it may assume that the ESCO would be unlikely to remove the lighting once installed. This is known as *opportunism*, 'self-interest seeking with guile' (Williamson, 1985, p. 4). A further interference with the efficient operation of transactions is *bounded rationality* as decision-making is subject to constraints on time, attention, resources and the ability to process information (Pint & Baldwin, 1997; Sorrell, 2005, 2007; Toffel, 2002) (see Table 4-1).

Table 4-1: Overview of characteristics and behaviours that can affect transaction costs (TC)

Affect	Competencies	Examples
Increase or reduce TC	Human capital specificity	Dedicated human resources for lighting
	Dedicated resources	Specialisation in a particular area
Increase TC	Bounded rationality	Limited time and capacity to analyse options
	Opportunism	One party taking selfish advantage of circumstances

Adapted from Pint and Baldwin (1997); Toffel (2002); Sorrell (2007).

Additional factors that influence the TCs of contracting are the *complexity of the task* (both replacing street lights or lamp posts and the complexity of monitoring performance according to contractual terms and conditions), the *competitiveness of the energy service market* and the *institutional context* in which contracting takes place (Sorrell, 2005). The institutional context can be affected by standardised tendering and procurement procedures and standardised M&V and projects such as the German 'LED Lead Market Initiative' (BMBF, 2014; SBI, 2013). Clients such as municipalities are only likely to enter an energy service contract such as an EPC for lighting if useful energy streams and final energy services can be supplied at a lower *total* cost compared to their provision using other modes of governance. Total cost are the sum of firstly PCs, *"the expenditures for inputs such as fuel and electricity"*, which depend on technical and operational efficiency of the equipment and secondly TCs as outlined in this chapter (Sorrell, 2007, p. 512).

To summarise, the viability of ESCO solutions is assumed to be determined by the following factors affecting TC (Pint & Baldwin, 1997; Sorrell, 2007; Toffel, 2002):

1. Technical asset specificity;
2. Human capital specificity;
3. Dedicated resources;
4. Task complexity;
5. Competitiveness of the market;
6. Institutional framework.

As we have identified the institutional framework as a factor also affecting the diffusion of EUEDs it is not included in our TC factors of the analysis in chapter 4.3.

4.3 Methods and data

We apply the TCE framework to help understand the relationship between the uptake of energy efficiency technology and ESCO solutions. Our analysis covers the intersection of LED and ESCO markets (where ESCO models are employed for LED diffusion) as a case study using a longitudinal analysis of key market players and key policy initiatives (Miles & Huberman, 1999; Patton, 2002).

4.3.1 Case study: LED street lighting and ESCO solutions in German municipalities

As pointed out earlier, LED is currently on the path towards maturity in the technology life cycle, although it still features high upfront costs, especially compared to conventional technologies. In Germany, LED sales are growing due to numerous initiatives at the federal level. Until 2013 the 'National Climate Initiative' provided subsidies to municipalities for switching to LED, which covered on average 30% of the costs. The 'LED Lead Market Initiative', founded in 2009, brings together market players and policy decision-makers to identify barriers for the municipal diffusion of LED. These activities take place in the context of a wider Eco-design context which phases out several energy-related products, such as many conventional lighting products, as a result of low efficiencies in 2015 (Schischke, Nissen, Stobbe, & Reichl, 2008).

9.5m public street lights installed in Germany consume approximately 4 TWh of electricity per year, which corresponds to energy costs of about 750 Mio €. This number represents approximately one third of municipal energy costs. The potential savings by using new energy efficient lighting systems (especially innovative LED) amount to 400 Mio. € per year (dena, 2012; DStGB, 2010), although numbers regarding refurbishments of old lighting systems are missing (dena, 2012). Germany's market for public street lighting is highly dispersed, consisting of more than 11000 municipalities with individual decision-makers.

Historically, different forms of governance for lighting provision have emerged. 27% of the municipalities provide street lighting in-house, 35% outsourced the management to EUCOs[12], another 10% to MUCOs[13] and 25% partially outsourced services such as maintenance. 3% of the municipalities use ESCO solutions to manage their street lighting (dena, 2012). Figure 4-2 depicts the main actors that take part in the process. Regulatory bodies oversee municipal finances and partially have to approve budgets, especially when the municipality is heavily indebted. Financiers mostly consist of (government-owned) banks.

Figure 4-2: Overview about actors and modes of lighting governance

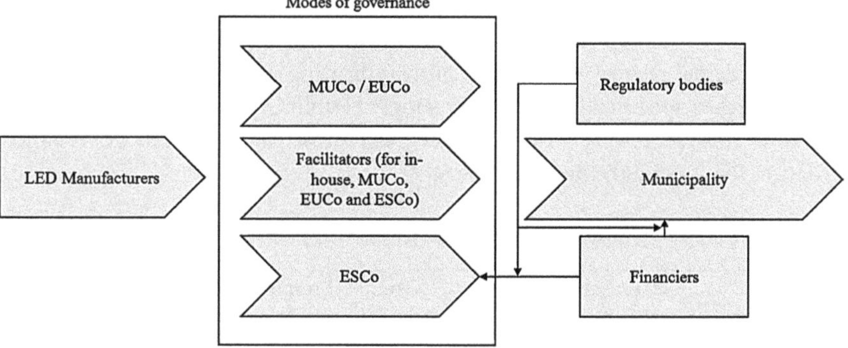

Despite the diversity of municipal lighting (and energy) governance in Germany, incumbent arrangements can represent a barrier for competition. The primary motivation for municipalities to engage with suppliers such as ESCOs is the

---

12 EUCOs stands for a national Energy Utility Company, which typically engage in energy generation and supply. They may also engage in distribution and transmission (Hannon, Foxon, & Gale, 2013, p. 1036).

13 MUCOs stands for Multi-Utility Company. In the German context this refers specifically to local 'Stadtwerke', which tend to provide a wide range of utilities such as gas, electricity and municipal waste management. Strong local embedding ensures that these MUCOs enjoy near monopolies on most supply and waste streams which can provide incentives for integrated solutions (Betsill & Bulkeley, 2006).

promise of final energy services such as lighting at a lower cost compared to in-house or MUCO provision.

The German ESCO market, one of the largest, most heterogeneous and most mature in Europe with approximately 250–500 active companies, experienced stable growth over recent years (Marino et al., 2011). ESCO solutions are gaining acceptance among customers although the market has reached only 10% of its potential (Duscha et al., 2013). We identified 10 companies that offer lighting services to municipalities, which include subsidiaries of EUCOs or infrastructure providers. Germany also has 4 EUCOs and around 900 MUCOs supplying energy (services) (Energy Transition, 2013).

As established in 4.2.4., the choice of supplier represents a choice of governance structure. In-house provision represents a hierarchical governance structure. EUCO contracts can be considered relational contracts as they focus on the terms of the relationship. MUCO contracts fall under the same category although their contractual relationship can be more dynamic as their business model may shift towards privatisation (market) or 'recommunalisation' (hierarchy). In fact, Germany is witnessing a trend towards 'recommunalisation' with 1/3 of municipal respondents to one study indicating interest in more hierarchical approaches specifically for energy supply (PublicGovernance, 2011). ESCO solutions represent long-term contracts, the most market-based governance structure for street lighting (see Table 4-2).

Table 4-2: Spectrum of governance structures for municipal street lighting in Germany

Vertical integration	—	**Relational contracts**	—	**Long-term contracts**	—	**Simple short-term contracts**	—	**Spot markets**
				Examples				
In-house lighting provision	—	MUCO or EUCO contract for lighting provision	—	ESCO solutions	—	Lighting arrangement for a recurring event	—	Lighting arrangement for a one-off event

Adapted from Pint and Baldwin (1997, p. 4).

ESCOs may be able to provide certainty over PC savings for a specified quantity and quality (performance) of lighting if replacement LED EUEDs can be financed, operated and maintained over the duration of a contract at lower cost compared to in-house or EUCO/MUCO provision. The complexity of long-term (energy performance) contracts compared to relational contracts or conventional in-house procurement of equipment increases the TCs of negotiating and managing the relationship with the (energy) service provider or manufacturer (combination of human capital specificity and bounded rationality). Consequently, PCs resulting from the physical characteristics of the energy system and the technical efficiency of organisational arrangements need to be lower for this model to be economically viable (Sorrell, 2007).

The case study of ESCO solutions for public sector LED street lighting provides a good example as potential savings are high (LEDs are ten times more efficient than halogen for the same light output (Bennich, Schoenen, Scholand, & Borg, 2014)) and the capacity of municipalities to invest in these EUEDs is, as mentioned above, limited (Hartmann et al., 2014). Municipal benefits of abandoning the in-house approach in favour of an ESCO solution may arise out of reductions in energy costs, less exposure to energy price fluctuation and the transfer of risk, allowing the municipality to focus on core activities. Outsourcing, however, also implies a loss of municipal control. There is also a danger of ESCOs cherry picking the most profitable municipal tenders to the detriment of less attractive ones. The viability of the EPC itself depends on the ESCO's ability to address technology factors, competency and capacity factors, as well as institutional factors which will be analysed in more detail in chapter 4.4.

4.3.2 Data Collection

Data was collected through an extensive longitudinal archival document analysis (i.e. context of lighting and ESCO markets from 2008–2013) and interviews, which permits us to reflect upon changes in the industry and institutional context. The documents consist of industry reports for lighting, energy services and public procurement compiled by industry experts such as energy agencies, official government bodies or industry associations. In addition as two of the three authors participated regularly in meetings of the German 'LED Lead Market Initiative', we also analysed the minutes of meetings regarding important topics for each of the participating stakeholders (i.e. manufacturers, government officials, ESCos and others).

The main empirical focus of this paper lies on the analysis of 40 semi-structured interviews with key stakeholders. The interviews took place from October 2013 until January 2014. We selected interview partners according to an approach suggested by Seawright and Gerring (2008). Following an initial screening and a consultation of experts from the 'LED Lead Market Initiative, we selected the most influential stakeholders in the process of LED application in German municipalities by that directly engage in the process of modernisation (see Figure 4-2). For each of the modes of governance we compiled typical cases that are representative. Hence we combined snowball sampling and purposeful sampling strategies. An overview of our sample can be drawn from Table 8-2. The interviews lasted 30–90 minutes and were conducted face-to-face or via telephone with one to two researchers present. The interviews have been recorded (for later verification) and notes have been taken. Representative quotes have then been translated into English. Questions during the interviews revolved around two main topics, notably modernisation of public street lighting (participating actors, processes and facilitators) and the role of ESCO solutions (see chapter 8.2.1 in the appendix for detailed questions).

4.3.3 Data Analysis

We systematically evaluated the collected archival documents from 2008 onwards. We then analysed the interview protocols and reflected these against the bulk of documentations. We analysed the material according to the research question concerning barriers to the uptake of innovative EUED technologies as well as the possible role of ESCO solutions for accelerated uptake of public LED street lighting. In this process we identified the following main topics:

- Technological drivers and barriers
- Institutional drivers and barriers
- Economic and financial drivers and barriers
- Competency and capacities as both are required to examine and technological, institutional and economic/financial barriers and to enhance the diffusion of innovative EUEDs

These were reflected against concepts of TCE (technical asset specificity, human capital specificity, dedicated resources, task complexity and competitiveness of market) in an 'abductive' process (Mantere, 2008), i.e. the back and forth between theory, interview transcripts and archival documents. We then compared the modes of lighting governance, highlighting the suitability for each in different situations (financial situation, competency and capacity of actors, institutional set-up) to derive policy implications for municipalities and government as well as management implications for innovative EUEDs.

4.4 Results: Factors affecting the diffusion of LED street lighting and EPC

This chapter provides an overview of factors affecting the uptake of EUEDs and the corresponding TCE interpretation. We present the drivers and barriers that increase or lower the TCs and PCs for different modes of governance (see Table 4-3 for an overview).

4.4.1 Technological factors

LED lighting provides a technological advantage over conventional lighting: higher energy efficiency. Despite a high potential for energy (and PC) savings, the first set of barriers to LED diffusion comprises technological aspects such as the maturity of the products, complexity and uncertainty regarding energy savings and lifetime. *'There is tremendous uncertainty regarding new measures to evaluate the performance of LEDs, for example maintenance factors and payback periods' (Manufacturer).* LED customers such as municipalities and ESCOs in particular further highlighted the missing standardisation and short warranties of LED due to the early stage in the innovation cycle as well as technological path-dependency relating to less innovative and less efficient lighting systems which are currently being installed. *'Many sales people from LED manufacturers have been to our town. There are too many standards and warranty mechanisms' (Municipal representative).* As this municipal representative suggests, technical change and the lack of standardisation may increase opportunism on behalf of the technology providers. Barriers associated with the technical asset specificity of LED lighting and the task complexity of both assessing technology offers up-front and M&V for potential savings translate into higher TCs.

4.4.2 Competency and capacity factors

A push for open-book accounting was a robust success factor for lowering TCs as the availability of consumptions figures lowers the risk of opportunism on behalf of the municipality since anticipated cost-savings are one of the main drivers for municipalities to modernise their lighting infrastructure. This lowers the human capital specificity for the assessment of alternative governance structures, specifically for EPCs as it reduces the ESCO's need for acquiring information to perform the contract. *I think one of the main barriers is the municipal accounting system. Usually the overall costs for running their lighting system are higher than what we estimate during the planning process (ESCO).*

The interviewees further highlighted the positive role of facilitators (e.g. energy agencies) and other consultants as well as best practice examples during the tender and implementation process as they provided the necessary lighting and planning expertise. Facilitators hold a critical position as they are able to reduce the risk of opportunism by making the task of determining cost savings less complex. These success factors lower the TC for any alternative governance structure as independent advice provides a more concrete overview of available options, which reduces search costs and the human capital specificity required to search for alternative options. If, on the other hand, intermediaries and facilitators between ESCOs and municipalities suffer from a lack of lighting competence, a bias towards established technologies and a lack of understanding about the complex tendering processes, the task of determining savings may remain complex. *The building and energy context require a lot of expertise. Energy consultants are lacking the competence for lighting functional tendering. So manufacturers and their tools to calculate savings dominate the market (Energy consultant).* Low human capital specificity of facilitators increases the danger of opportunism by suppliers, which increases TCs for ESCO solutions.

Cost transparency and cost management systems facilitate outsourcing solutions for modernisation as the procuring agency can compare baseline and future scenarios. *The main preconditions for effectively modernising the lighting systems lie in a lighting database and a clear guideline. So you turn that into a functional, neutral tender (Municipal representative).* In addition, the early involvement of decision-makers from local government and administration further facilitate their use. Neutral tenders ensure technology-free bidding and realisation, which are more likely to be an ESCO competence. Manufacturers enter into competition for the collaboration with an ESCO, which results in lower prices and higher quality lighting products, which then can be implemented in ESCO projects. These factors reduce human capital specificity and task complexity as potential savings can be anticipated and adequate tenders be set up, thus lowering TCs for ESCOs as services provided by the contract are well defined.

Competency barriers relate to the capacity of the actors to overcome technological, market and institutional barriers. *Many people simply ignore the risks and do the modernisation on their own (ESCO).* On the one hand, municipal representatives often fail to evaluate the market for LED lighting concerning quality, energy savings and risks. *The implementation of the modernisation is demanding. You need to know the technical details and test examples in practice (Municipal representative).* On the other hand, municipal representatives often lack the administrative competency to design tenders to reflect appropriately their quality and endurance criteria and carry out comprehensive budgeting and cost management throughout the procurement process of modern LED street lighting. *Usually [municipal representatives] don't know about their own costs for lighting. Data about the old lighting systems in terms of energy costs, investments etc. is missing (Manufacturer).* These competencies were not necessary in the past, as more efficient technologies evolved slowly. In the case of low human capital specificity dedicated towards municipal energy management, outsourcing may be a sensible option but the difficulty of designing tenders may prove too complex a task for a municipality to consider, translating into high TCs and a difficult choice between realisation of modernisation and possible outsourcing options. Our results show that the responsible decision-makers tend to ignore or reject these solutions as the perceived human capital specificity is too high.

4.4.3 Institutional factors

From the analysis we derive institutional drivers which have a varied effect on the TCs involved for different governance modes to modernise municipal lighting. These include an alignment of interests between different administrative bodies that are responsible for lighting, acceptance for the new lighting technology by the local population and the procuring organisation. Additionally, financial service providers act as drivers of energy efficiency projects, since they provide necessary risk and return structures.

Barriers on the demand side relate to institutional problems such as the property situation (many municipalities sold their street lighting to national EUCOs) and specific structural arrangements for the provision of public street lighting which has historically often been a task for MUCOs. *Existing contracts with national EUCOs often run for a very long time and the national EUCO only complies with the legal minimum when it comes to efficiency. A switch to LEDs, which would make sense, does not happen. They use less efficient technologies (ESCO).* The interviewees also referred to the inadequate infrastructure for LED technology. Time consuming and costly procedures to switch to another contractor translate into human capital specificity and increase TCs for ESCO solutions as a favourable institutional context is needed.

Experts also referred to problems on the supply side (i.e. in the ESCO market) as only few ESCOs exist that target the lighting market. *Street lighting does not receive the attention it deserves in terms of potential cost savings and improvements in lighting quality (Energy agency).* Manufacturers, which could also act as ESCOs did not show willingness to enter the service based market segment as they perceived the margins as low and complexity as high despite providing secondary conversion equipment. Specialised ESCOs gain a competitive advantage in the field of lighting, although the low competitiveness in the ESCO market increases TCs as it reduces the likelihood of municipalities considering outsourcing through EPCs.

When, on the other hand, municipalities are willing to engage in ESCO solutions, experiences tend to be positive. *The risks taken by the ESCO exceed the amount of financial savings I have when I do the modernisation on my own (Municipal representative).* In this case human capital specificity associated with lowering the risk of in-house EUED retrofit is too high, which reduces TCs for long-term contracts and more market-based governance structures.

Efforts required to govern the relationship relate to the task complexity and human capital specificity of negotiating and monitoring contracts. *There is a complexity problem. Many contract documents are longer than 50 pages (Financial service provider).* Regarding EPC design, the experts highlight transparency, comprehensibility and a distinct guarantee for energy savings as beneficial. *Guidelines, transparency of the contract and an ESCO that selects high quality products turned out to be successful (Municipal representative).* Increasing know-how and enabling structures for the modernisation process could significantly reduce task complexity of ESCO solutions particularly if contractual arrangements align interests by providing transparency and flexibility.

Regarding the EPC design, municipal actors in particular also emphasised the need for flexibility (e.g. to change or improve lighting systems) and a fair balance of interests during the contract negotiations. *We need flexibility regarding, short-medium- and long-term developments in the markets (Municipal representative).* An exact definition of the baseline, however, often proves difficult. *Complexity of EPC leads to high TCs. Exact numbers are needed (Energy agency).* EPCs also need to be checked by the regulatory bodies when municipalities run on a very tight budget in need of consolidation. *The regulating authorities need to approve the EPC. The financing over the contract duration needs to be assured (Municipal agency).* Administrative approval represents a dedicated institutional resource to mitigate against bounded rationality on behalf of the municipality.

Additionally, long-term partnerships with manufacturers are often in place, which create technological and product-related lock-in effects. Manufacturers possess a lot of market power. They dictate the tenders because they have long lasting partnerships and thus can charge individual prices for each customer including ESCOs, MUCOs or EUCOs. *The manufacturers are the winners in this market. They often have long lasting partnerships with the municipalities, they supply analyses for free. Many tenders thus specify one product (ESCO).* Contractual barriers also relate to lighting arrangements with national EUCOs, which prevent ESCOs from entering the public lighting market. *The contracts [with national EUCOs] run for a very long time. The municipal representatives do not have expertise any more (Energy agency).*

4.4.4 Economic / investment factors

Economic drivers result from PC savings, which EUCOs and MUCOs can realise by lowering the cost for energy and better procurement conditions thanks to long lasting contracts with manufacturers. However, these companies might be subject to a conflict of interests, as they engage in selling electricity as opposed to providing energy efficiency services. Conservative attitudes towards innovative technologies may also prevent them from deploying LED. ESCOs, on the other hand, possess better procurement conditions for lighting equipment, which potentially favours an outsourcing solution.

Economic barriers stem from the relatively high upfront costs of LED street lighting, particularly for municipalities as they typically run on a tight budget, in some cases tightly controlled by regulatory bodies with budgetary control powers (see Table 4-3). Volatile energy prices and uncertain price developments for LED further constrain the payback of investments. These factors prevent municipalities from realising significant production cost savings, a prerequisite for considering any alternative mode of governance for modernisation.

EUCOs and MUCOs possess knowledge about the current lighting system and experience in providing maintenance services for the municipality. Hence a EUCO/MUCO solution may exhibit lower aggregate PCs for the provision of the final energy services. *ESCOs need to refinance themselves. However, forfeiting is not accepted by many municipalities. This leads to financial constraints for ESCOs (Financial service provider).* Hence an ESCO solution might exhibit higher financing costs, which reduces its potential to cut PC.

Table 4-3: TCE perspective – Factors affecting the governance of novel EUEDs

Factors affecting uptake of LED and choice of governance model		Transaction cost perspective	Actions and human behaviour	Specific TCs for governance model	
				Lowering TCs	Increasing TCs
Technological factors	*Measurement and verification of savings*	Technical asset specificity and task complexity	Increases opportunism on behalf of the manufacturer		For ESCO solution
	Lack of standardisation	Technical asset specificity and task complexity	Increases opportunism on behalf of the manufacturer		For general outsourcing
	Uncertainty regarding warranties	Technical asset specificity and task complexity	Increases opportunism on behalf of the manufacturer		For general outsourcing
Competency and capacity factors	*Open-book accounting*	Human capital specificity	Reduces opportunism on behalf of the municipality	For ESCO solution	
	Expert facilitators	Human capital specificity	Reduces opportunism on behalf of the municipality	For ESCO solution	
	Cost transparency and neutral tenders	Human capital specificity and task complexity	Reduces time-consuming and costly search for alternatives	For ESCO solution	
Institutional factors	*Lock-in contracts with existing suppliers*	Human capital specificity	Time-consuming and costly search for alternatives to in-house, MUCO or EUCO solution		For ESCO solution
	Low ESCO market competitiveness	Competitiveness of the market	Time-consuming and costly search for alternatives to in-house, MUCO or EUCO solution	For in-house solution	For ESCO solution
	Risk transferral	Human capital specificity	Risk taken by ESCO surpasses financial savings of in-house provision	For ESCO solution	

Factors affecting uptake of LED and choice of governance model		Transaction cost perspective	Actions and human behaviour	Specific TCs for governance model	
				Lowering TCs	Increasing TCs
	Transparency of outsourcing procedure and contracts	Human capital specificity and task complexity	Transparency enables municipalities to consider ESCO solution	For ESCO solution	
	Administrative approval procedure	Dedicated resources	Mitigates against bounded rationality on behalf of the municipality	For in-house solution	
Economic/ investment factors	*EUCOs and MUCOs potentially selling and saving energy*	Competitiveness of the market	ESCO independence from selling energy reduces conflict of interest	For in-house or ESCO solution	
			Allows ESCOs and MUCOs to reduce overall PCs by reducing electricity tariffs for lighting	For EUCO or MUCO solution	
	Volatile energy prices and uncertain technological development trajectories	Human capital specificity and task complexity	Time-consuming and costly search for alternatives		For in-house solution
	Experience of current lighting system and providing maintenance	Dedicated resources	ESCOs lacking experience	For EUCO or MUCO solution	

4.5 Discussion

4.5.1 Factors affecting the diffusion of LED street lighting and associated services

Previous research points towards technological, economic, competency and institutional factors influencing the diffusion of innovative EUEDs (Klein Woolthuis et al., 2005; Rogers, 1995; Schleich, 2009; Sorrell et al., 2004; Wilson et al., 2012) and we analyse each in turn with regard to public sector EUED application.

First, our results show that technological factors can represent particularly strong barriers in the case of public sector EUED application due to specific institutional set-ups / regulations and lower competency compared to private actors, which limits the public sector ability to carry out a mission-oriented procuring policy or tenders (Edler & Georghiou, 2007; Edquist & Zabala-Iturriagagoitia, 2012; Uyarra et al., 2014). This relates directly to the second set of factors, namely competencies and capacities. Building on previous work (Jacobsson & Karltorp, 2013; Klein Woolthuis et al., 2005; Schleich, 2009; Sorrell et al., 2004), we highlight their absence to scope and analyse markets for innovative technologies and possible ways to source them as the most important factor hindering the widespread adoption of EUEDs. Third, our results also show that at the interplay between services and technologies, institutional factors such as established relationships, flexibility and balance of interests are crucial to advance the uptake of innovative EUEDs (Maxwell et al., 2006; Maxwell & van der Vorst, 2003). Hence the interactions between procurer and innovator are critical to address the increasing complexity in procuring services and infrastructures (Hartmann et al., 2014; Roehrich & Lewis, 2014; Testa et al., 2014). Fourth, economic barriers to ESCO solutions, such as the low priority of energy efficiency projects, uncertainty, legal and institutional problems, access to finance and difficulties defining baseline consumption figures (see Pätäri and Sinkkonen (2014); Marino et al. (2011) and Zhang et al., (2014)) prevail throughout our case study. Our assessment of technological, institutional and economic barriers points towards enhanced competencies on the municipal side (see Table 4-3) as a prerequisite for considering alternative governance structures that may accelerate the diffusion of innovative EUEDs.

4.5.2 Modes of governance for novel end-use energy demand technologies

Technological, economic and competency barriers to EUED diffusion can be addressed through a change in governance structure. Institutional factors enabling transparency by providing dedicated capacities, such as independent energy agencies, enable alternative governance structures to be considered in the first place. In the absence of in-house capacities to assess alternative options, intermediary organisations play important roles in reducing opportunism and enhancing transparency. This is particularly relevant in situations where customers 'wait' for further technological improvements (Roehrich et al., 2014; Schleich, 2009; Sorrell et al., 2004), instead of actively engaging in the diffusion process. The best mode of governance for EUED diffusion therefore depends on the institutional set-up in which the municipality is embedded and the capacities it can dedicate towards the task.

First of all, municipalities could manage the modernisation of the street lighting installations in-house. This requires significant technical and commercial know-how to evaluate technologies, markets and the institutional background to achieve an economically viable solution. These translate into high TCs for modernisation as the human capital and technical asset specificity associated with potentially risky investments in innovative technology are high. There is also a danger of lower PC savings by deploying inefficient products. On the other hand, if sufficient capacities are available and risk can be adequately mitigated against, in-house solutions provide the greatest opportunities for cost savings by maximising municipal control.

Second, municipalities in Germany have historically tended to choose relational contracts with EUCOs or MUCOs for municipal street lighting. As these trusted relationships feature interwoven knowledge of existing technologies and infrastructure (Backlund & Eidenskog, 2013; Hannon et al., 2013), the task complexity for contract management and the human and technical asset specificity of EUCO or MUCO solutions for modernisation are likely to be low (see Table 4-3). Relational contracts particularly with a MUCO could therefore have an advantage over a third-party ESCO solution. On the other hand, these companies usually supply energy to the municipality as well, which reduces their incentive to apply innovative EUED technologies.

Third, with an ESCO solution, customers may achieve cost savings from the beginning of the contract and additional cost savings due to freed personnel capacity. However, the transfer of risks and uncertainty regarding technological components and development goes hand in hand with a loss of municipal decision-making rights. One representative of a facilitating organisation precisely framed the potential role of ESCOs: '*[Energy performance] contracting is a means to accelerate the diffusion and adoption of innovative energy efficiency technologies.*' We thereby confirm earlier research suggesting that PPPs can foster innovation, technical knowledge and skills as well as managerial efficiency from a purely technological perspective (Painuly, Park, Lee, & Noh, 2003; Roehrich et al., 2014; Testa et al., 2014).

We extend the view of previous studies (Hypko et al., 2010a; Hypko, Tilebein, & Gleich, 2010b) that performance-based contracts can lead to lower costs in the provision of energy services, although customers demand high flexibility during the contract duration as they fear a lock-in into unknown new technologies. In that case, scholars suggest an emphasis on the establishment of trusting relationships, particularly during the set-up of the contract (Backlund & Eidenskog, 2013; Roehrich & Lewis, 2014), which may again benefit a MUCO solution. To fully exploit the potential of ESCO solutions in the municipal context, we can confirm the need for tenders not to over-specify to allow technology neutral bidding originally proposed by Uyarra et al. (2014).

4.5.3 Using TCE for the analysis of ESCO solutions

TCE enables the analysis of governance structures for EUED diffusion by comparing drivers and barriers that constitute technological, economic, institutional or competency factors and how these in turn relate to higher or lower TCs. Many of our case study participants considered the energy service fee of ESCO solutions comparatively low compared to the risk and complexity associated with in-house search for new products and services. In the absence of MUCOs and trusted EUCOs, ESCOs can therefore reduce human capital specificity required to perform modernisation in-house and contribute to reducing opportunistic behaviour towards the municipality, especially if ESCOs act as facilitators between lighting manufacturers and municipalities. Table 4-3 suggests that standard procurement rules, model contracts as well as facilitators could help lower TCs for ESCO solutions, which coincides with earlier findings by Bleyl et al. (2013) and Roehrich and Lewis (2014).

Above all, our major contribution revolves around the TCE perspective on the uptake of innovative EUEDs and the different modes of governance. Factors influencing the diffusion of EUEDs (technology, competency, institutional and economic) have been related to factors influencing the TCs of different governance modes (technical asset and human capital specificity, dedicated resources, task complexity, competitiveness), taking into account human behaviours relevant to TCs (opportunism and bounded rationality) to compare the viability of ESCO solutions vis-a-vis other modes of governance. Public sector LED street lighting is likely to fulfil several preconditions hypothesized by Sorrell (2007) for outsourcing using EPCs, although some assumptions about municipal street lighting do not hold as competency and institutional problems prevail.

Our findings indicate that EPCs might even be a means to reduce occurring TC and thus accelerate the diffusion and application of EUEDs, especially when a trusted MUCO is absent and the municipality features low in-house competency and a tight budget. Based on our results we argue to include the notion of competencies and capacities (quality and quantity) which tend to be 'hidden' behind 'dedicated resources' and 'human capital specificity' in TCE frameworks (Pint & Baldwin, 1997; Sorrell, 2007; Toffel, 2002).

4.6 Conclusions and implications

4.6.1 Conclusions

An accelerated uptake of innovative EUEDs among public actors such as municipalities is economically and socially desirable as it bears a huge (mostly untapped) potential for reducing carbon emissions and energy dependency and providing a relief for public budgets. It may also increase economic growth in the (domestic) EUEDs industry.

Our results show that municipalities can often choose from a range of municipal street lighting governance options. Where this is not the case, ESCOs represent a vehicle for the commercialisation and diffusion of innovative EUEDs by addressing barriers to technological diffusion through EPCs. This could be a PPP when procuring innovative products in a public context. In this regard two key conclusions can be drawn: First, the design and content of tenders emerged as a key phase during the adoption of LED technologies and the assessment of governance options. To address competency and capacity barriers, criteria to design these tenders should be widely diffused in order to increase competition among organisational (governance) structures, including documentation and guidelines or instruments to calculate baseline and savings (e.g. with standard tendering processes). Second, intermediaries (e.g. consultants or energy agencies) emerged as a key driver for modernisation in general and ESCO solutions in particular as they also address competency and capacity gaps, which represent the main barrier.

On the other hand, high associated TCs might hinder the diffusion of the ESCO solution and need to be addressed specifically. Therefore TCE has proven as a useful analytical tool. We also highlight the discrepancy between theoretical fit of ESCO solutions and actual diffusion. As mentioned above, only 3% of the municipalities use this mode of governance, which points towards the existence of strong interdependencies between actors and a conservative institutional environment. This 'lock-in' may be a reason for the slow diffusion of LED lighting and/or ESCO solutions. Our study therefore also points towards an institutional environment in Germany that appears to favour relational contracts over long-term contracts. The presence of a MUCO in a German municipality provides a particularly strong case for maintaining such a set-up as it provides opportunities to strengthen local capacity to manage energy and benefit from the diffusion of innovative EUEDs, rather than outsourcing.

4.6.2 Implications for policy makers and managers

In terms of policy and managerial implications we want to highlight the benefits of market transparency regarding both novel EUEDs and ways to source them.

On the policy level, introducing statutory obligations for tendering to include outsourcing options can benefit the choice among governance models. The same holds true for technological standards or enhanced warranties to address technological complexity as well as standard contracts to address legal complexity. These could be established at a national level. Additionally, policy-makers need to address institutional barriers by providing the infrastructure necessary to facilitate the transition towards innovative EUEDs for example by rethinking long-term partnerships with MUCOs, EUCOs or other partners such as lighting manufacturers at the municipal level. Depending on the institutional set-up, establishing an EPC with a MUCO could also be a way to lower the costs and risk associated with modernisation. As highlighted above, to further foster the ESCO market, policy makers need to support facilitators (e.g. consultants or energy agencies) to disseminate specific technological and commercial knowledge. This could be done either by subsidising consultancy services or by supporting the market for energy consulting by establishing remunerations schemes to provide incentives for consultants to enter the market. EPC for lighting also faces a lack of finance, which represents a significant economic barrier. Government owned banks (such as Germany's KfW) or governments themselves could provide credit guarantees to increase the access to finance to increase the competitiveness of the LED street lighting market as a policy response on the national level.

On the managerial level, ESCOs targeting the public market for energy efficiency should focus on a combination of products and services that fits the underlying prolonged technological lifecycle. Hence providing municipalities with a range of complementary services would increase their customer loyalty and reduce energy consumption and related costs in the long run. ESCOs need to follow an open-book approach in order to attract attention from municipal clients demanding high transparency, flexibility and a balance of interests. The use of standard contracts might be a means to meet these requests (SBI, 2013).

4.6.3 Limitations and future research

Our use of TCE in exploratory qualitative research, as opposed to rigorous model testing, exhibits limitations regarding the measurability of the constructs. We nevertheless believe that our analysis uses TCE appropriately. Conclusions derived from this study, however, do not represent generalizable assumptions about the usefulness of ESCO solutions and specifically EPCs for the diffusion of EUEDs as this process is usually not linear. Many factors other than those analysed in this paper influence the diffusion of innovative EUEDs and need to be taken into account (e.g. behavioural factors) to establish the appropriateness of ESCO solutions.

Additional markets could be valuable to explore as the viability of ESCO solutions in different institutional contexts. Finally, the conditions for ESCOs to maintain a competitive position vis-à-vis other actors (EUCOs, MUCOs, manufacturers, etc.) provide an interesting field of future research (Hannon et al., 2013). In relation to that, scholars could further explore why the creation of a market for EPC in a public context did not achieve its potential so far (Pätäri & Sinkkonen, 2014).

Acknowledgements

The authors are grateful for the time and support of the interview participants as well as Florian Täube (Solvay School of Management and Economics) and conference reviewers for valuable comments on earlier versions of the paper. The opportunity to present and discuss earlier versions of the paper at ETH PhD Academy 2014 and the Energy Systems Conference 2014 helped us tremendously in further refining our argument. The research team would like to thank the German Federal Ministry of Education and Research (BMBF) for their financial support as part of the research project "OLPOMETS" and the Centre on Innovation and Energy Demand, SPRU, University of Sussex funded by the RC UK's EUED Programme (grant number EP/KO11790/1).

5 Public policy influence on renewable energy investments – a panel data study across OECD countries

Friedemann Polzin, Michael Migendt, Florian A. Täube & Paschen von Flotow

Abstract: This paper examines the impact of public policy measures on renewable energy (RE) investments in electricity-generating capacity made by institutional investors. Using a novel combination of datasets and a longitudinal research design, we investigate the influence of different policy measures in a sample of OECD countries to suggest an effective policy mix which could tackle failures in the market for clean energy. The results call for technology-specific policies which take into account actual market conditions and technology maturity. To improve the conditions for institutional investments, advisable policy instruments include economic and fiscal incentives such as feed-in tariffs (FIT), especially for less mature technologies. Additionally, market-based instruments such as greenhouse gas (GHG) emission trading systems for mature technologies should be included. These policy measures directly impact the risk and return structure of RE projects. Supplementing these with regulatory measures such as codes and standards (e.g. RPS) and long-term strategic planning could further strengthen the context for RE investments.

5.1 Introduction

Climate change has been aggravating in recent years as CO_2 emissions continue to grow in developed and developing countries alike (IPCC, 2014). OECD countries have a larger responsibility to address climate change, as they accounted for almost 50% of the global carbon emissions in 2010, most of which were linked to the energy sector (BNEF, 2013; Müller, Brown, & Ölz, 2011; OECD, 2013). One option for transitioning towards a low-carbon society is to increase the share of energy (especially electrical energy) generated from renewable energy (RE) sources (Foxon & Pearson, 2007; Jefferson, 2008). To complement these activities, it is necessary to also increase overall efficiency of the energy sector (Eichhammer, Ragwitz, & Schlomann, 2013; Marques & Fuinhas, 2011; Schleich, 2009), however throughout this article we focus on capacity additions of RE. Given limited financial resources of state governments, the involvement of private markets and investors is needed

111

(Mathews et al., 2010). Investments in clean energy[14] projects and infrastructure have been growing in the last decade and total up to a relevant amount. Yet the volumes are still smaller than investments in conventional fossil fuel-based power (see Figure 5-1) (Frankfurt School-UNEP Centre & BNEF, 2014).

Figure 5-1: Renewable power capacity investment compared to fossil-fuel power capacity investment, 2008–2013 in billion USD

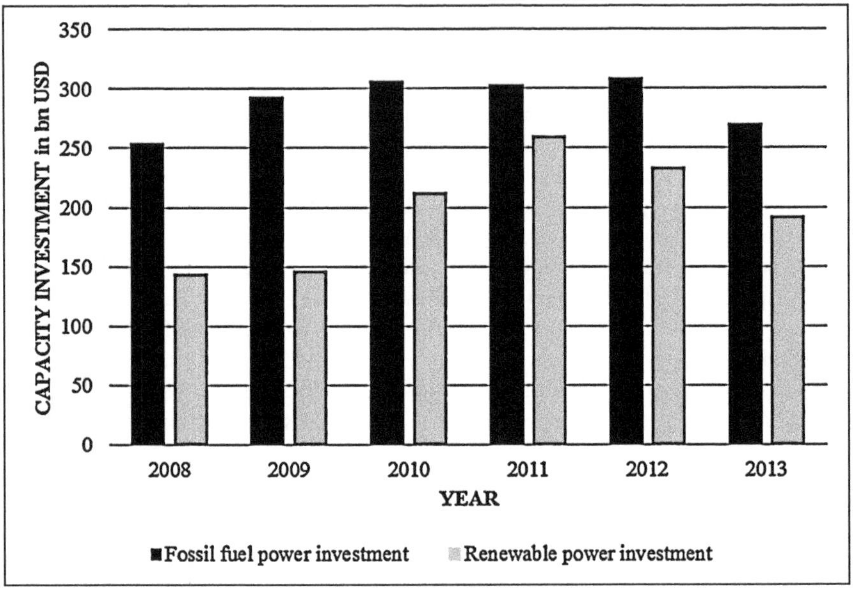

Adapted from Frankfurt School-UNEP Centre and BNEF (2014, p. 31)

This is partially due to the challenge the diffusion of RE technology presents (Foxon & Pearson, 2007; Friebe, Flotow, & Täube, 2013; Friebe et al., 2014; Veugelers, 2011). Market failures occur related to inherent characteristics of the energy sector which is designed for fossil fuel-based power plants on the one hand, as well as the particular nature of RE technologies on the other (Dinica, 2006; Helm, 2002; Jefferson, 2008; Wüstenhagen & Menichetti, 2012). Long pay-back periods and illiquid assets coupled with high regulatory dependencies and corresponding uncertainties often make RE unattractive or even unsuitable for

---

14 Throughout this paper, we will use *renewable energy (RE)* and *clean energy* inter-changeably.

investors. While this holds also true for mature technologies and conventional power plants, REs in addition face some technological uncertainty (Kenney & Hargadon, 2012; Mathews et al., 2010).

To mitigate market failures and to compensate for technological and economic weaknesses, policy makers were historically confronted with a variety of options to stimulate diffusion of RE. This article attempts to uncover the effects of different policies to support clean energy applications. It adds to recent academic and political discussions about the choice among feed-in tariffs (FITs) and alternative mechanisms as well as the corresponding overall effectiveness and efficiency of these policy instruments (Carley, 2009; Lesser & Su, 2008; Mathews et al., 2010). However quantitative analyses of the influence of public policies on investments in RE by private *institutional* investors (i.e. investment and pension funds, banks and insurance companies) are scarce, with longitudinal, let alone panel data analyses almost non-existent (Bolkesjø, Eltvig, & Nygaard, 2014; Chassot et al., 2014; Lüthi & Prässler, 2011; Lüthi & Wüstenhagen, 2012). This is even more relevant because private institutional investors provide a significant amount of the funds deployed, particularly in certain more mature technologies such as wind (BNEF, 2013; IEA, 2013b).

Thus, building on previous work on RE investor behaviour (Bergek et al., 2013; Chassot et al., 2014; Lüthi & Prässler, 2011; Wüstenhagen & Menichetti, 2012), we investigate the influence of public policies on subsequent RE investments by institutional investors across OECD countries over a time period of 12 years (2000–2011). Using a novel dataset gathered from Bloomberg New Energy Finance (BNEF) and the International Energy Agency (IEA) together with the International Renewable Energy Agency (IRENA), this paper applies a panel data regression. We develop policy implications adding to the academic literature of RE policies and RE investment decisions. The overarching research question is: *Which policies have proven (most) conducive to investments in renewable energy assets?*

The remainder of this paper is structured as follows: Chapter 5.2 briefly describes the conceptual motivation. Chapter 5.3 introduces the analytical method and data used. Chapter 5.4 presents the results and discussion while chapter 5.5 finalises with conclusion, limitations and next steps in the research process.

5.2 Theoretical background

5.2.1 Public policy influence on renewable energy deployment

Literature on energy policy has analysed the relationship between policy and RE deployment in a number of different ways, generating evidence for policy makers to support their decisions (Delmas & Montes-Sancho, 2011; Harmelink, Voogt, & Cremer, 2006; Jacobsson et al., 2009; Menz & Vachon, 2006). Fundamental questions such as influence of price or quantity-based mechanisms have been addressed (Menanteau et al., 2003). Prior research thus provides a differentiated picture to support RE deployment.

First of all, fiscal and financial incentives can be provided to accelerate the development of RE projects. A number of scholars (Bolkesjø et al., 2014; Couture & Gagnon, 2010; De Jager et al., 2011; del Río & Bleda, 2012; Jenner, Groba, & Indvik, 2013; Lesser & Su, 2008; Mitchell et al., 2006) underline the superiority of feed-in tariffs (FIT) to spur deployment and technological diversity and lower risks for private actors associated with RE technologies. The implementation of FIT and changes to the remuneration system may harm the positive effects as seen for example in Spain. These positive effects hold when comparing FIT with other remuneration models such as quotas or auction-based systems (Butler & Neuhoff, 2008; Mitchell et al., 2006). However, Río and Bleda (2012) argue that a variety of policies consisting of specific and technology-neutral measures provide the most fertile ground for RE technology deployment.

To provide short term fiscal relief for RE projects grants and subsidies can be provided. In the logic of financing innovation this type of support instrument proves especially relevant in the early stages of technology development (Olmos et al., 2012) However it also reduces investors overall cost for RE projects (Bergek et al., 2013; De Jager et al., 2011; Olmos et al., 2012).

Similarly, government loans or loan guarantees could be of interest to private actors as the ability to refinance their activities is crucial for a long-term commitment in RE (Bergek et al., 2013; De Jager et al., 2008, 2011). A growing stream of literature has further analysed the impact of tax-based incentives to spur RE deployment (Barradale, 2010; Bird et al., 2005; Cansino, Pablo-Romero, Román, & Yñiguez, 2010; Quirion, 2010). However, Barradale (2010) highlights the missing policy commitment (due to direct dependency on the public budget) as a main shortcoming of taxes and rebates.

Second, scholars argue that in the sense of carbon and energy market liberalisation preference should be given to market-based instruments as first best solutions, e.g. carbon cap and trading systems (Helm, 2002; Rogge & Hoffmann, 2010; Rogge et al., 2011; S. Smith & Swierzbinski, 2007). Another market-based system, i.e. the tradability of RE certificates (green certificates) might further spur the deployment of renewables (Jensen & Skytte, 2002; Szabó & Jäger-Waldau, 2008).

Third, policy makers have the option to provide funds to local authorities to be spent on RE deployment (i.e. funds to subnational governments) (Bird et al., 2005; De Jager et al., 2008; Menz & Vachon, 2006; Ragwitz et al., 2008) or directly

114

invest in complementary assets such as infrastructure (De Jager et al., 2008, 2011; Henriot, 2013; Steinbach, 2013).

Fourth, investment decisions can be related to policy instruments that do not directly impact the risk and return structure of RE projects. For example, the perception of investment opportunities and preference for short term or long-term incentives also influence the decision to invest (Masini & Menichetti, 2012). Prior literature has found that the creation of the surrounding institutions is a major driver to facilitate investments in RE technologies (Bergek et al., 2013; Wüstenhagen & Menichetti, 2012). This could be accompanied by a long-term strategic framework which is valued by institutional investors as they prefer stability in cash flows over the duration of their investments (De Jager et al., 2008, 2011; Lüthi & Wüstenhagen, 2012), thus foreseeable changes to regulations and policy consistency are paramount (White, Lunnan, Nybakk, & Kulisic, 2013).

Finally, regulatory measures to stimulate markets need to be established even though diffusion and application of RE technologies is socially and politically desirable (Jefferson, 2008). Furthermore, the highly regulated environment for diffusion of mature REs might require new forms of regulation compared to the commercialisation of RE in order to overcome market failures and dissolve path dependencies. Research has on the one hand highlighted mixed effects of renewable portfolio standards (RPS) for the US (Delmas & Montes-Sancho, 2011; Carley, 2009; Bird et al., 2005). Carley (2009), Bird et al. (2005) and Menz & Vachon (2006) show that RPS systems increase the share of RE produced but not the absolute amount, whereas Delmas and Montes-Sancho (2011) do not find a significant effect on RE capacity. On the other hand, further mandatory requirements, quota, and obligation schemes do exhibit a positive influence on RE application (Menz & Vachon, 2006); however, they have proven inferior to other instruments such as FIT (Butler & Neuhoff, 2008; De Jager et al., 2008, 2011; Mitchell et al., 2006).

One solution to address the problems encountered so far, apparently, is a 'policy mix' consisting of complementary instruments. However, no scholarly consensus exists on what the optimal policy mix could look like (Foxon & Pearson, 2007) or on which criteria should be applied to determine it (Carley, 2009). For example, Del Rio and Bleda (2012) argue that a variety of policies, consisting of technology-specific and technology-neutral measures is needed to enhance deployment of RE technologies.

Following this debate scholars applied quantitative statistical methods (i.e. panel data analysis on country level) to investigate policy impacts on RE deployment. For example, Johnstone et al. (2010) found that market based approaches favour technologies that are close to be technologically and cost wise competitive with fossil fuels (such as wind) whereas feed-in tariffs are conducive to innovation in less mature technologies (such as solar). On the other hand, Popp et al. (2011) do not find a significant effect of either FIT or a renewable certificate system on wind power investments.

Building on this work, scholars tried to uncover the influence of a number of different policy instruments on the contribution of renewables to the total energy supply (Marques & Fuinhas, 2012a). They show that aggregated measures such as fiscal and financial incentives (including FIT) as well as measures, that seek to define strategies and outline specific programs to promote these RE sources, have a positive significant impact. Controlling for a range of political elements such as energy security, Kyoto protocol ratification and socio-economic factors (e.g. prices for fossil-fuels, welfare) Aguirre and Ibikunle (2014) found no significant positive influence of policies on RE growth, however they found a negative contribution of fiscal and financial incentives (i.e. taxes).

In sum, the literature has been searching for an integral overview of sustainable energy policy to spur RE diffusion. Given the mixed results of prior findings, we add to this debate through our research which aims at uncovering the effectiveness of different disaggregated policy instruments (i.e. the optimal policy mix) to induce private finance in RE assets. More specifically, we contribute to two different discussions: The FIT versus other support scheme debate, and the overall assessment of different support mechanisms for RE.

Beyond economic, regulatory and behavioural barriers, institutional factors such as acceptance among the local communities hinder the deployment of RE technologies (Arabatzis & Myronidis, 2011; Tampakis, Tsantopoulos, Arabatzis, & Rerras, 2013; Wüstenhagen, Wolsink, & Bürer, 2007). With our analysis we contribute to the private sector perspective on RE investments, leaving aside potential negative externalities caused by technologies deployed (Friebe et al., 2014).

5.2.2 Investors' perspective on renewable energies

The question of how to effectively mobilise financial resources for the deployment of RE (RE projects) and complementary infrastructure has been a major concern both in the academic and political debate (Bergek et al., 2013; De Jager et al., 2011; Mathews et al., 2010; Mowery et al., 2010; Veugelers, 2011; Wüstenhagen & Menichetti, 2012). Mathews et al. (2010, p. 3263) note that 'the issue of public vs. private financing is not yet adequately explored', but add that there is consensus among policy makers that the transition to a low carbon economy will not happen without the involvement of private institutional investors (Müller et al., 2011; Popp, Hascic, et al., 2011).

Investments in RE deployment by institutional investors (i.e. investment/pension funds, banks and insurance companies) are typically hindered by a number of factors; high upfront costs, risks, and uncertainty regarding long-term viability of the technology, long payback periods, high regulatory and infrastructural dependency as well as public acceptance (Cárdenas-Rodríguez, Johnstone, Haščič, Silva, & Ferey, 2013; De Jager et al., 2011; Haley & Schuler, 2011; Kenney & Hargadon, 2012; Müller et al., 2011). These factors directly influence the risk/return profile of an RE investment, which is a major determinant for institutional investors (Cárdenas-Rodríguez et al., 2013; Dinica, 2006).

The ultimate requirement for a sustainable RE policy is a reduction of capital costs to create a level playing field with fossil fuel-based technologies which have been heavily subsidised in the past (Szabó & Jäger-Waldau, 2008). To lower these costs properly, policy makers have to take into the account factors that influence institutional investors. Scholars analysed the decision criteria for investors. Bergek et al. (2013) suggest evaluation criteria such as (overall or portfolio) cost, perceived (market) uncertainty and political risk. Based on insights from project developers, regulatory risks and the streamlining of the administrative process (grid access) have been identified as relevant decision criteria (Friebe et al., 2014; Lüthi & Prässler, 2011; Lüthi & Wüstenhagen, 2012). Chassot et al. (2014) confirm these propositions by highlighting the perceived risk caused by policies as the main determinant of investment decisions.

Szabó & Jäger-Waldau (2008) suggest that a more competitive financial environment could actually reduce the costs of capital for RE projects, given that capital markets function efficiently. Therefore, a combination of supportive financial regulation and transparent policy making would be conducive to institutional investors to compete for building RE capacities (Bergek et al., 2013; Lüthi & Wüstenhagen, 2012; Wüstenhagen & Menichetti, 2012). This environment would in turn lower the perceived regulatory risk and, thus, lower the financing costs for RE projects while still allowing reasonable rates of return. Decreasing support corresponding to technological development, meaning adaptive policy making, can furthermore spur the deployment of more innovative technologies (Szabó & Jäger-Waldau, 2008).

With our analysis we explore which policy instruments have been conducive to RE investments by institutional investors, thus we examine this relationship over time. We answer the call for research of Wüstenhagen & Menichetti (2012) by specifically investigating the influence of policy on the perception of institutional investors.

5.3 Methods and data

5.3.1 Research design

Investigating the diffusion of a particular technology and corresponding investments requires a longitudinal research design (Angrist & Pischke, 2008; Cárdenas-Rodríguez et al., 2013; Johnstone et al., 2010; Popp, Hascic, et al., 2011; Wooldridge et al., 2009). Prior literature applied panel data regressions at the EU level (Bolkesjø et al., 2014; Marques & Fuinhas, 2012a) with a few comparing OECD or BRIC countries (Aguirre & Ibikunle, 2014; Cárdenas-Rodríguez et al., 2013; Johnstone et al., 2010; Popp, Hascic, et al., 2011).

Building on these methodological approaches we cover a variety of OECD countries, conducting a panel data regression throughout the time period from 2000 to 2011 to explain the influence of policy instruments on the diffusion of clean energy technologies. As policy instruments do not exhibit an immediate effect on technology application and investments, we add a lag procedure which is explained in chapter 5.3.3 (Wooldridge et al., 2009). The time frame is chosen due to the limited availability of high quality data, and as it still covers the most substantial developments in the worldwide renewable energy sector, especially regarding the involvement of institutional investors.[15] Globally the wind sector grew from 18 GW installed capacity in 2000 to 238 GW installed capacity in 2011, while the solar sector grew from 1.5 GW installed capacity in 2000 to 67 GW installed capacity in 2011 (IEA, 2014c). The biomass sector, at the same time, grew from 38 GW installed capacity in 2000 to 74 GW installed capacity in 2011 (NREL, 2013).

5.3.2 Data

Data was collected from two independent data sources. Investments (additions in RE capacity) have been drawn from BNEF, which possesses one of the most comprehensive databases in the field of clean technology financing (Cárdenas-Rodríguez et al., 2013). It contains information on installed electricity generating capacity, date, transaction type, financing type, amount of equity, and amount of debt. We used data from 2003 to 2011. The complete database includes 5840

---

15 When using data from BNEF, one has to be aware of the fact that the data is being updated after a certain year. Thus we have chosen to limit the data of the dependent variable to 2011, as this timeframe covers the most reliable data quality.

Solar Investments, 9643 Wind Investments and 2889 Biomass & Waste Investments. These three RE subsectors account for 72% of the RE funds invested and with 74 GW installed for 75% of all capacity additions in the period, and have therefore been selected for further analysis. The records of the BNEF database start with the year 2003 and we further decided to use the data up to the year 2011, as data quality of the following year did not meet the standards of previous years at the time of research in early 2013.

Policy indicators were drawn from the IEA/IRENA Policy and Measures (PM) database which includes policy measures in OECD countries from 1974 onwards. These indicators have been used by scholars to analyse the impacts of aggregated policy instruments in Europe (Cárdenas-Rodríguez et al., 2013; Marques & Fuinhas, 2012a, 2012b) and globally (Aguirre & Ibikunle, 2014). We used the IEA/IRENA policy measures data from 2000–2011 which corresponds to the availability of investment data. The database includes a total of 957 distinct policy measures active in the respective years resulting in 7835 policy data points (see chapter 8.3.2 in the appendix).

We structured the data according to sectors (Multiple RE sources, Wind, Solar, and Biomass) and applied additional data processing: First, we removed cases with missing values and included only completed deals. Second, our interest in the influence of policy measures lead to the exclusion of countries with less than three consecutive years of investor activities. The case-selection is carried out for each sector analysed (see chapter 5.3.3. for the procedure and Table 8-3 for a list of countries in the subsamples). The selection of policy instruments included in the model is described in chapter 5.3.3.

As control variables we further included macroeconomic and energy market data from the US Energy Administration – EIA (carbon intensity of the economy), OECD (long-term interest rates, share prices) and from the World Bank (total energy consumption, gross domestic product (GDP)) in our analysis.

5.3.3 Model

We investigate the influence of the set of different policy measures on subsequent investments into RE capacity by institutional investors. The components of this model are shown in Figure 5-2. The five different subsets of policies include 957 different (countries and time series generated) measures of which some are active in every sector, while some are active in a single sector only. The included variables are log-transformed to correct for the skewed distribution of both dependent and independent variables (Hair, 2010). We expect to see a positive influence for every subset of policies on the capacity additions (see Figure 5-2).

Figure 5-2: Model for the quantitative panel regression

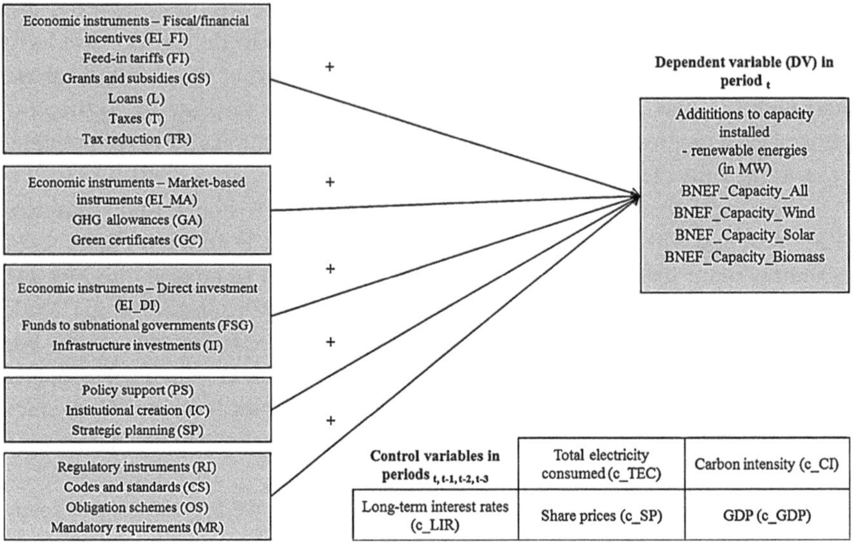

Control variables in periods $_{t,\,t-1,\,t-2,\,t-3}$	Total electricity consumed (c_TEC)	Carbon intensity (c_CI)
Long-term interest rates (c_LIR)	Share prices (c_SP)	GDP (c_GDP)

Dependent variable (DV)

The dependent variable, drawn from BNEF, for the overall model is measured as aggregated newly installed capacity (in MW) in a certain country and year in a specific subsector (e.g. solar, wind, biomass). We use capacity indicators as they represent the most accurate proxy for the deployment of a technology (Popp, Hascic, et al., 2011).

Independent variables (IV)

The main independent variables were drawn from the IEA/IRENA database. We constructed our indicators by counting distinct policies per country per year (see Table 8-4 for a complete categorisation). They are measured by the number of active instances of policies affecting the RE sector (Johnstone et al., 2010; Marques & Fuinhas, 2012a). We call these counts 'accumulated number of RE policies and measures' (ANPM) (Aguirre & Ibikunle, 2014). The active policy instances are used as an ordinal variable with the assumption that the more policies (of a kind) the better for investments. This allows comparing the experience of many countries and decomposing the effect of distinct factors econometrically (Cárdenas-Rodríguez et al., 2013).

We can further distinguish between technology specific instruments (e.g. specific targets for certain energy sources) and instruments that apply to all types of RE (e.g. German FIT). The IEA/IRENA PM database provides relevant information on characteristics (title, country, year (started and ended), policy status (e.g. in force, ended, superseded), policy type, policy target (e.g. subsector such as solar, wind and biomass), geographical scope (supranational, national, regional), policy sector (e.g. electricity, multi-sectoral, framework policy), size of plant targeted (large, small or both) and funding (partially, depending on instrument)). We build our model with distinct policies as independent variables based on prior literature reviewed in chapter 5.2.

Controls

To account for technological progress, economies of scale as well as the fact that the installed capacity gains momentum (leading to variance from the previous years), we include year and country dummy variables. To rule out alternative explanation for RE investments we included a number of control variables in the regressions. Economies of scale are picked up by the time dummy variable and therefore not included among the control variables. Further economic indicators that might drive capacity additions include the GDP (c_GDP). To account for differences in energy use and consumption we include energy dependency (CO_2 intensity – Metric Tons of Carbon Dioxide per thousand year 2005 U.S. dollars GDP) as well as electricity consumption in the regression (c_CI, c_TEC). Finally, to account for factors influencing investor behaviour, we include interest rates (c_LIR) as well as share prices (c_SP) of local indices as these might render an investment into RE vs. non-RE more or less attractive.

Lag structure

As common for longitudinal analyses we assume that the independent variable affects the dependent variable immediately and with a certain delay. This relation is included in our model through a lag structure where we compare the respective time values of the DV with several different time values of the IV. This approach helps in this case to account for the time-dependent influence of policy measures on investor behaviour (Angrist & Pischke, 2008; Wooldridge et al., 2009). We decided to introduce a lag structure with a lag of the policy effectiveness within zero to three years (i.e. investments in capacity in year $_t$ are influenced by policy measures in years $_{t,\,t-1,\,t-2,\,t-3}$). This means policies could have a direct effect on investor decisions or it could take up to three years for a policy to trigger a capacity addition. On the one hand, it is possible that investors anticipate the regulation and already have their projects ready when it is passed, as the regulatory process is mostly open. On the other hand, there are factors delaying the investment process such as the time needed to build the wind farm or solar park and to gain access to the grid. In general, changes in regulations evolve throughout time and are communicated prior to them being passed and set active.

5.3.4 Longitudinal analysis (panel data regression)

Prior research acknowledges that determining the influence of policy measures on investments in RE capacity is challenging since spatial and temporal effects could overlap (Marques & Fuinhas, 2012a; Marques, Fuinhas, & Pires Manso, 2010). Following their work we assume panel auto-correlation and contemporaneous correlations as similarities in policy design (e.g. in EU countries) and a tendency to increase the number of policy measures which can be observed throughout the data. The literature reviewed in chapter 5.2 analyses policy measures that are conducive to RE investments, which are included in the model (see chapter 5.3.3). The best fitting econometric technique is to use a panel data approach under the conditions resulting from characteristics of policy making (Marques & Fuinhas, 2012a).

We estimate panel corrected standard error (PCSE), ordinary least squares (OLS) and random effects estimator (REE) models. Following Marques and Fuinhas (2012a) and to mitigate errors resulting from the data structure, we use several econometric treatments: heteroskedasticity, panel autocorrelation, and contemporaneous correlation are addressed through fitting approaches (Reed & Ye, 2011). Thus, we circumvent inefficient coefficient estimation and a biased estimation of standard errors (Marques & Fuinhas, 2012a). We do not focus on OLS or REE as they do not address serial correlation and contemporaneous correlation, however, we include the estimates for reasons of robustness (Marques & Fuinhas, 2012a, 2012b).

Our analysis proceeds as suggested by Marques and Fuinhas (2012a): 1. We observe the quality and nature of the data; 2. We test the presence of heteroskedasticity, panel autocorrelation and contemporaneous correlation; 3. If the results do not conform to standard assumptions about errors (i.e. if the error terms are Independent and identically distributed – iid), we employ the PCSE estimator, which is a suitable solution to improve the accuracy of the estimators; 4. We compare the results with those derived from OLS and REE to check the robustness.

Table 5-1: Specification tests for the quantitative model

	Random effects / Pooled OLS				Fixed effects			
	Multiple RE	**Solar**	**Wind**	**Biomass**	**Multiple RE**	**Solar**	**Wind**	**Biomass**
Modified Wald test	-	-	-	-	236.84***	91.24***	1052.72***	141.05***
Wooldridge test F(N(0,1))	19.20*** (OLS)	12.63*** (OLS)	19.62*** (OLS)	1.22 (OLS)	-	-	-	-
Pesaran's test	19.55***	5.54***	16.26***	8.70***	15.02***	4.13***	10.15***	4.61***
Frees' test	3.00***	1.03***	2.49***	0.90***	2.34***	0.22	1.87***	0.63**
Friedman's test	78.00***	30.63***	63.77***	38.51***	25.35**		45.38**	27.18*

Notes: The Wald test has a Chi^2 distribution and tests the null hypothesis that none of the independent variables are significant; The Wooldridge test is N(0,1) distributed and tests the null hypothesis that there is no serial correlation. Pesaran and Frees' tests examine the null hypothesis that there is cross-sectional independence; Pesaran's test is a parametric procedure which follows a standard normal distribution; Frees' test employs Frees' Q-distribution; Friedman's test is a non-parametric estimation based on Spearman's rank correlation coefficient (Aguirre and Ibikunle, 2014; Hoyos and Sarafidis, 2006; Marques and Fuinhas, 2012a). ***, **, *, denote 1, 5 and 10% significance level, respectively. xtcsd and xtserial commands were used.

Table 5-1 presents results from the estimations of the specification tests regarding quality and nature of the data and confirms that especially the policy data is heteroskedastic (i.e. has a common variance) and that panel autocorrelation and contemporaneous correlation is present. Hence we use PCSE estimator as main econometric analysis technique.

As public policy effects differ across RE subsectors (e.g. solar, wind, biomass) we carried out the further analysis sector by sector and also aggregated the data (Multiple RE sources) to analyse effects that are similar across sectors. Thereby, we can also distinguish policy instruments between the sectors as well as policies that apply to all sectors.

Panel data estimation without lag procedure (I)

$$IC_{jk} = const + \sum_{i=1}^{i} \beta_i PM_{ijk} + C_{jk} + d_j + d_k + \varepsilon_{jk}$$

Panel data estimation with lag procedure (II–IV)

$$IC_{jk} = const + \sum_{i=1}^{i} \beta_i PM_{ijk-l} + C_{jk-l} + d_j + d_{k-l} + \varepsilon_{jk-l}$$

IC_{jk} is the aggregated installed capacity financed by institutional investors per country j per year k. PM_{ijk} is a vector of i explanatory variables representing policy measures based on the IEA/IRENA scheme (per country per year). C_{jk} consists of a number of control variables (Model I). For the analyses of time-dependent phenomenon we include lags l of one to three years in the regressions (Model II–IV). The dummy variables d_j and d_k refer to country and time, respectively. The PCSE estimator permits the error term ε_{jk} to be correlated over the cases (i.e. countries). Moreover a first-order autoregression for ε_{jk} over time can be used. Finally the estirmator allows ε_{jk} to be heteroskedastic (Cameron & Trivedi, 2009; Marques & Fuinhas, 2012a).

5.4 Results and discussion

The aim of our research is to uncover the influence of different policy instruments on subsequent investments in RE by private institutional investors over time in a longitudinal research design. Policy makers interested in improving their country's transition towards RE should implement measures for attracting private institutional investors, as the capital required for large-scale RE projects by far surpasses the available funds of utility companies as well as the public budgets. Institutional investors' capital played an important role in the development of the RE sector, and establishing a favourable environment for them, including specific policies, should increase capacity additions in the future. With our analysis we provide an integral picture of RE policies and their influence on RE capacity investments by institutional investors. We intend to contribute to the literature surrounding investor behaviour regarding RE technologies, as investors provide funds for large scale deployment (Bergek et al., 2013; Lüthi & Prässler, 2011; Wüstenhagen & Menichetti, 2012).

The analysis is conducted on a sectoral basis to allow differentiated policy recommendations. In the following discussion, we highlight significant effective and ineffective policy measures and relate our results to previous studies in this literature stream (Aguirre & Ibikunle, 2014; Marques & Fuinhas, 2012a; Marques et al., 2010).

Table 8-4 shows the descriptive statistics of our analysis. The correlation among explanatory variables has been subject to analysis as well. The simultaneous use of several drivers leads to the hypothesis of collinearity among explanatory variables. Table 8-4 and Table 8-5 show the summary statistics and the correlation coefficients for our analysis. The analysis suggests the absence of collinearity among the exogenous (independent) variables.

Table 5-2: Panel-corrected standard errors (PCSE) regression results

Independent variables (ANPM) $_{(t-1)}$	PCSE							
	No autocorrelation		No autocorrelation		No autocorrelation		No autocorrelation	
	Multiple RE		Wind		Solar		Biomass	
	(I)		(II)		(III)		(IV)	
	Coeff.	S.E.	Coeff.	S.E.	Coeff.	S.E.	Coeff.	S.E.
$EI\_FI\_FI_{jk}$	0.69***	0.28	0.70***	0.26	1.18***	0.26	0.09	0.24
$EI\_FI\_GS_{jk}$	0.33	0.25	0.13	0.24	0.61***	0.23	0.61**	0.24
$EI\_FI\_L_{jk}$	−0.36	0.26	0.30	0.27	−0.91***	0.30	−0.16	0.32
$EI\_FI\_TR_{jk}$	0.40**	0.18	0.20	0.20	−0.01	0.28	0.10	0.18
$EI\_FI\_T_{jk}$	−0.90*	0.48	−0.51	0.34	1.00**	0.45	−0.20	0.29
$EI\_MI\_GA_{jk}$	1.48**	0.62	1.59***	0.41	−1.17**	0.60	0.97*	0.47
$EI\_MI\_GC_{jk}$	−0.02	0.34	−0.03	0.35	−2.27***	0.41	0.06	0.34
$EI\_DI\_FSG_{jk}$	−0.65*	0.40	−0.38	0.43	−1.43***	0.45	1.50***	0.39
$EI\_DI\_II_{jk}$	0.15	0.26	0.37	0.24	0.21	0.40	−1.20**	0.39
PS_IC_{jk}	−0.41	0.28	0.13	0.26	−1.51***	0.38	0.83**	0.37
PS_SP_{jk}	0.70*	0.37	0.16	0.43	2.35***	0.30	−0.44*	0.24
RI_CS_{jk}	0.45***	0.17	0.63**	0.19	0.54**	0.27	−0.41	0.37
RI_OS_{jk}	0.28	0.26	−0.09	0.31	0.16	0.31	−0.01	0.27
RI_MR_{jk}	0.52	0.36	0.21	0.32	0.77**	0.37	0.28	0.27
Control variables								
c_TEC_{jk}	0.64	0.22	0.75**	0.36	0.95	0.74	−1.97***	0.65
c_CI_{jk}	−0.09	0.67	1.94**	0.94	−6.01***	2.24	4.37***	1.56
c_LIR_{jk}	−0.67***	0.19	−0.66***	0.19	−0.54	0.75	−0.57**	0.29
c_SP_{jk}	1.84***	0.56	1.58***	0.62	−0.24	0.59	0.61	0.42
c_GDP_{jk}	−0.02	0.08	−0.06	0.08	−0.82	0.70	2.16***	0.58
_cons	−7.12**	3.29	−6.74**	3.47	21.59	17.18	−53.24***	13.25
Observations	330		319		176		220	
R²	0.38		0.40		0.49		0.38	
Wald	258.81***		208.87***		161.69***		169.40***	

Notes: The Wald test has a Chi^2 distribution and tests the null hypothesis of non-significance of all coefficients of independent variables; panel corrected standard errors are reported. ***, **, *, denote significance at 1, 5 and 10% significance levels, respectively; Estimates include country and time dummies (Marques and Fuinhas, 2012a). xtpcse command was used.

125

We estimated all models separately using the PCSE and the OLS estimator as well as REE for robustness checks. We conducted the analysis for Multiple RE data and distinct sectors. The estimation results are displayed in order of the categories and different policy measures. We report the results with models based on a time lag of 1 year, thus, investments lagging behind the introduction of policy by 1 period. With this analysis we go beyond extant work (Aguirre & Ibikunle, 2014; Johnstone et al., 2010; Marques & Fuinhas, 2012a), ruling out reverse causality (i.e. investments driving policies for example through lobbying) and providing a more realistic approach to renewable deployment, taking into account a lagging reaction of investors to policy measures. The results of our complete policy variable analysis are presented in Table 5-2 and in Table 5-3. An overview about our results can be drawn from Table 5-4.

Table 5-3: Ordinary least square (OLS) regression results

Independent variables (ANPM) $_{(t-1)}$	OLS							
	Standard errors		Standard errors		Standard errors		Standard errors	
	Multiple RE		Wind		Solar		Biomass	
	(V)		(VI)		(VII)		(VIII)	
	Coeff.	S.E.	Coeff.	S.E.	Coeff.	S.E.	Coeff.	S.E.
$EI\_FI\_FI_{jk}$	0.69**	0.30	0.70***	0.26	1.18***	0.31	0.09	0.26
$EI\_FI\_GS_{jk}$	0.33	0.26	0.13	0.25	0.61**	0.25	0.61**	0.25
$EI\_FI\_L_{jk}$	−0.36	0.44	0.30	0.40	−0.91**	0.43	−0.15	0.39
$EI\_FI\_TR_{jk}$	0.40	0.33	0.20	0.29	−0.01	0.34	0.10	0.27
$EI\_FI\_T_{jk}$	−0.90*	0.49	−0.51	0.44	1.00*	0.58	−0.20	0.39
$EI\_MI\_GA_{jk}$	1.48**	0.67	1.59***	0.61	−1.17	0.80	0.97	0.53
$EI\_MI\_GC_{jk}$	−0.02	0.44	−0.03	0.41	−2.27***	0.48	0.06	0.37
$EI\_DI\_FSG_{jk}$	−0.65	0.52	−0.38	0.47	−1.43***	0.47	1.51***	0.52
$EI\_DI\_II_{jk}$	0.15	0.44	0.37	0.37	0.21	0.56	−1.20***	0.46
PS_IC_{jk}	−0.41	0.41	0.13	0.34	−1.51	0.43	0.83**	0.36
PS_SP_{jk}	0.70**	0.35	0.16	0.36	2.35***	0.34	−0.44	0.30
RI_CS_{jk}	0.45	0.33	0.63**	0.32	0.54	0.36	−0.41	0.33
RI_OS_{jk}	0.28	0.39	−0.09	0.35	0.16	0.41	−0.01	0.34
RI_MR_{jk}	0.52	0.37	0.21	0.34	0.77*	0.47	0.28	0.31

Independent variables (ANPM)$_{(t-1)}$	OLS							
	Standard errors		Standard errors		Standard errors		Standard errors	
	Multiple RE		Wind		Solar		Biomass	
	(V)		(VI)		(VII)		(VIII)	
	Coeff.	S.E.	Coeff.	S.E.	Coeff.	S.E.	Coeff.	S.E.
Control variables								
c_TEC$_{jk}$	0.64***	0.19	0.77***	0.20	0.95	0.72	−1.97***	0.52
c_CI$_{jk}$	−0.09	0.96	0.96	1.02	−6.01**	2.58	4.37***	1.60
c_LIR$_{jk}$	−0.67***	0.24	−0.74***	0.24	−0.54	0.67	−0.57*	0.30
c_SP$_{jk}$	1.84***	0.33	2.10***	0.35	−0.24	0.46	0.61*	0.37
c_GDP$_{jk}$	−0.02	0.10	−0.05	0.09	−0.82	0.70	2.16***	0.50
_cons	−7.12***	2.65	−8.33***	2.59	21.59	17.01	−53.24***	11.51
Observations	330		319		176		220	
R2	0.38		0.40		0.49		0.38	
F	10.05***		10.93***		8.01***		6.52***	
Mean VIF	2.03		1.94		9.17		5.15	

Notes: The F-test is normally distributed $N(0,1)$ and tests the null hypothesis of non-significance of the coefficient estimates taken together. ***, **, *, denote significance at 1, 5 and 10% significance levels, Estimations include both country and time dummies (Marques et al., 2010). regress command was used.

5.4.1 Fiscal and financial incentives

First of all, our results highlight the effectiveness of FIT to spur capacity additions which directly impact the risk and return structure of RE projects as a policy instrument that guarantees a certain return on investment and provides an incentive for investors. FIT have been implemented in a range of countries, starting with Germany and Austria in 2000. The aggregated results (Multiple RE) as well as wind and solar sector results revealed a highly significant positive coefficient. However, the effect differs across sectors. Whereas in the solar sector FIT has a stronger impact than overall, FIT is less effective in the wind sector.

FITs proved particularly successful in countries such as Germany and Italy with some exceptions in other countries (e.g. Spain) (Bolkesjø et al., 2014; Couture & Gagnon, 2010; del Río & Bleda, 2012; Jenner et al., 2013; Lesser & Su, 2008; Mitchell et al., 2006). This instrument is a strong signal to investors as it addresses the capital market restrictions by adjusting the risk/return structure (Cárdenas-Rodríguez et al., 2013). Thus our research is in line with evidence by Río and Bleda (2012) who underline the superiority of FITs to spur deployment and to lower risks associated with RE technologies. In addition our research confirms that a variety of policies consisting of specific and technology-neutral measures spur RE technology deployment.

Second, our results show that grants and subsidies prove to be effective as short term measures to alleviate finance constraints. This holds true for the solar and biomass sectors. Grants and subsidies temporally reduce the cost of finance for a project, and directly depend on a public budget (Johnstone et al., 2010) which makes them more unstable than for example FIT. This is shown by our results of the solar sector. Institutional investors exhibit a preference of FIT over subsidies. With this analysis we confirm earlier work, that highlighted this type of support instrument can also contribute during the diffusion stages of the innovation cycle (Bergek et al., 2013; Bolkesjø et al., 2014; Olmos et al., 2012).

Third, we provide evidence for the effectiveness of loans and loan guarantee programs. According to our results this type of instrument does not spur the deployment in the solar sector. This stands in contrast to prior literature which showed that loans and loan guarantees could enhance institutional investors ability to refinance by reducing cost of capital (Bergek et al., 2013; De Jager et al., 2008, 2011)

Our research regarding tax-based measures revealed mixed evidence. On the one hand aggregated results point towards a negative effect of taxes on subsequent capacity additions in renewables; however results from the solar sector reveal a positive impact. On the other hand, tax reductions tend to increase overall capacity in renewables, with no particular effect in the sectors. We therefore confirm earlier research that pointed towards an ambivalent nature of taxes as they depend on public budgets (Barradale, 2010; Cansino et al., 2010; Quirion, 2010).

Above all we confirm (Marques & Fuinhas, 2012a) who found that fiscal and financial incentives are conducive to increasing the share of RE. We extend their results by decomposing the indicator into distinct policy measures. On the other hand, we contrast and extend Aguirre and Ibikunle (2014) who found a negative effect of fiscal and financial incentives. We disentangle this view, showing that FIT and grants and subsidies can have positive effects; however, we do find a negative effect for taxes.

Institutional investors' decisions are supported through direct influence on their return side of their investment calculations. Higher income through grants and subsidies and lower capital costs through FITs support their openness towards RE investments (Lüthi & Wüstenhagen, 2012). Tax regulation does not necessarily have conducive effects as many institutional investors already have a tax optimised corporate structure.

5.4.2 Market-based incentives

Our analysis also provides mixed evidence for market based incentives such as emission trading schemes and green certificate schemes on subsequent RE investments. GHG emission trading systems have been introduced in the US, Australia, UK, Italy and Norway since 1991. First, aggregated results as well as sectoral results from the wind and biomass sectors highlight the robust positive influence of tradable permit systems. This might be on the one hand due to the fact that wind energy can be installed in relevant capacities. On the other hand biomass plants generate a constant flow of certificate an exhibit base-load characteristics. We thereby confirm earlier literature on GHG systems (Helm, 2002; Quirion, 2010; Rogge & Hoffmann, 2010; Rogge et al., 2011; S. Smith & Swierzbinski, 2007).

Interestingly, the presence of GHG emissions allowances have a stronger impact on the capacity financed by institutional investors than FIT, as investors prefer market-based systems which are less dependent on policy changes which is the case in aggregated results and the wind and biomass sectors. Our results add empirical evidence to the debate revolving around FITs and tradable permits (Butler & Neuhoff, 2008; Cansino et al., 2010).

On the other hand, GHG emissions allowances show a negative impact on the capacity financed by institutional investors in the solar sector. Solar energy technologies, being less mature, cost-effective and more diverse than wind, are more heavily dependent on regulation, although grid parity is almost reached. These developments are reflected in our results regarding the policy mix. Market based incentives (such as GHG emission trading systems) prove to be ineffective in this case. A possible explanation lies in the fact, that solar technologies require stronger signals towards investors due to their relative novelty compared to wind technologies. We thereby confirm Johnstone et al. (2010) who found that market based approaches favour technologies (such as wind) that are closer to competitiveness with fossil fuels whereas feed-in tariffs are conducive to innovation in less mature technologies (such as solar).

Second, green certificates that permit trading the obligatory RE capacity in a national scheme do not incentivise institutional investors to install more solar capacity which might be due to the lack in maturity and the low amount of certificates generated per capital invested compared to other sectors, such as wind. These measures depend on the total quantity in the market which might vary resulting in insufficient mobilisation of funds and high regulatory uncertainty. In addition, quota-based systems tend to be opaque as they involve over-the-counter transactions for certificates (Cárdenas-Rodríguez et al., 2013). Our results contrast Szabó and Jäger-Waldau (2008) as well as Jensen and Skytte (2002) who conceptualised a positive influence of these renewable certificate markets. In sum, institutional investors like the openness of market based policy measures, as long as they promise reliable support for their investments (Chassot et al., 2014). Possible risks through technological or natural uncertainties lead to hard to calculate returns which lower the appetite for investments into RE.

Table 5-4: Overview of the results

Independent variables (ANPM) $_{(t-1)}$	Overview							
	Multiple RE		Wind		Solar		Biomass	
	PCSE	OLS	PCSE	OLS	PCSE	OLS	PCSE	OLS
$EI\_FI\_FI_{jk}$	+ (***)	+ (**)	+ (**)	+ (***)	+ (***)	+ (***)	+ (NS)	+ (NS)
$EI\_FI\_GS_{jk}$	+ (NS)	+ (NS)	+ (NS)	+ (NS)	+ (***)	+ (**)	+ (**)	+ (**)
$EI\_FI\_L_{jk}$	– (NS)	– (NS)	+ (NS)	+ (NS)	– (***)	– (**)	– (NS)	– (NS)
$EI\_FI\_TR_{jk}$	+ (**)	+ (NS)	+ (NS)	+ (NS)	– (NS)	– (NS)	+ (NS)	+ (NS)
$EI\_FI\_T_{jk}$	– (*)	– (*)	– (NS)	– (NS)	+ (**)	+ (*)	– (NS)	– (NS)
$EI\_MI\_GA_{jk}$	+ (**)	+ (**)	+ (***)	+ (***)	– (**)	– (NS)	+ (*)	+ (NS)
$EI\_MI\_GC_{jk}$	– (NS)	– (NS)	– (NS)	– (NS)	– (***)	– (***)	+ (NS)	+ (NS)
$EI\_DI\_FSG_{jk}$	– (*)	– (NS)	– (NS)	– (NS)	– (***)	– (***)	+ (***)	+ (***)
$EI\_DI\_II_{jk}$	+ (NS)	+ (NS)	+ (NS)	+ (NS)	+ (NS)	+ (NS)	– (**)	– (***)
PS_IC_{jk}	– (NS)	– (NS)	+ (NS)	+ (NS)	– (***)	– (NS)	+ (**)	+ (**)
PS_SP_{jk}	+ (*)	+ (**)	+ (NS)	+ (NS)	+ (***)	+ (***)	– (*)	– (NS)
RI_CS_{jk}	+ (***)	+ (NS)	+ (**)	+ (**)	+ (**)	+ (NS)	– (NS)	– (NS)
RI_OS_{jk}	+ (NS)	+ (NS)	– (NS)	– (NS)	+ (NS)	+ (NS)	– (NS)	– (NS)
RI_MR_{jk}	+ (NS)	+ (NS)	+ (NS)	+ (NS)	+ (**)	+ (*)	+ (NS)	+ (NS)
Control variables								
c_TEC_{jk}	+ (NS)	+ (***)	+ (***)	+ (***)	+ (NS)	+ (NS)	– (***)	– (***)
c_CI_{jk}	– (NS)	– (NS)	+ (**)	+ (NS)	– (***)	– (**)	+ (***)	+ (***)
c_LIR_{jk}	– (***)	– (***)	– (***)	– (***)	– (NS)	– (NS)	– (**)	– (*)
c_SP_{jk}	+ (***)	+ (***)	+ (***)	+ (***)	– (NS)	– (NS)	+ (NS)	+ (*)
c_GDP_{jk}	– (NS)	– (NS)	– (NS)	– (NS)	– (NS)	– (NS)	+ (***)	+ (***)
_cons	– (**)	– (***)	– (**)	– (***)	+ (NS)	+ (NS)	– (***)	– (***)
R^2	0.38	0.38	0.40	0.40	0.49	0.49	0.38	0.38

***, **, *, denote significance at 1, 5 and 10% significance levels, respectively; NS refers to 'not statistically significant'.

5.4.3 Direct investments

According to our results from the biomass sector positive contributions with high significance include funds to sub-national governments (direct investments of federal money with regional, local or municipal level entities as intermediaries or targets) as biomass markets tend to be regionally dispersed. We hereby confirm earlier works (based on wind sector analyses) which highlighted that this form of direct investment spurs RE deployment (Bird et al., 2005; De Jager et al., 2008; Menz & Vachon, 2006; Ragwitz et al., 2008). Our results from the solar sector revealed that this type of instrument is ineffective in promoting solar capacity addition. This might be due to the fact that institutional investors ignore subsidies on a local level.

Infrastructure investments to provide grid access seem ineffective for channelling investor's money into biomass technologies. This stands in contrast to previous literature which highlights the grid expansion as conducive to RE deployment and investment (De Jager et al., 2011; Henriot, 2013; Steinbach, 2013). Reasons for deviating results in the biomass energy sector (compared to solar and wind) lie in the different structure which can be characterised by a strong regional focus and usually the small scale of power plants (Upreti, 2004). This might attract a different set of investors that focus less on overall market conditions.

5.4.4 Policy support

When examining the influence of policy support measures on subsequent investments into RE capacity, our results show interesting characteristics. First, institutional creation, such as the implementation of an energy agency, accelerates the capacity additions in the biomass sector effectively. Results from the solar sector show the contrary. A possible explanation lies in the fact that the sudden cut in the Spanish feed-in tariff system is coded as an institutional creation/change.

Secondly, aggregated and solar sector results show that a clear long-term energy strategy (strategic planning) is conducive to investments as investors favour a long-term framework with a clear vision. Essentially, all countries in our study had policy measures containing a strategic component, however only a few incorporated it in many policy initiatives.

With this analysis, we add empirical support for the strong role which a long-term policy commitment (strategic planning) plays in an effective policy mix. De Jager et al. (2011) state that commitment, stability, reliability and predictability are all elements that increase confidence of market actors, reduce regulatory risks, and hence significantly reduce cost of capital. Additional evidence points towards a preference for policy consistency even when changes occur (White et al., 2013). Thus, our results confirm conceptual and empirical works by several scholars, which hold clear strategic long-term economic instruments to be conducive to RE investments (Lüthi & Wüstenhagen, 2012; White et al., 2013; Wüstenhagen & Menichetti, 2012). Finally, we confirm work by Marques and Fuinhas (2012a) who highlight the policy processes and strategy frameworks as positive for RE deployment. Possible risks through instability in regulation directly affect investors' return calculations and determine investment decisions immediately (Wüstenhagen & Menichetti, 2012). Therefore, a long-term vision for the regulatory environment through strategic planning and most likely through institutional creation will support the investment environment for institutional investors.

5.4.5 Regulatory instruments

Regulatory mechanisms and institutionalisation of markets in the form of codes and standards (especially RPS) also attract institutional investors. According to our results this is the case for the aggregated sectors as well as in the wind sector, perhaps because the wind sector shows elements of a developed market based on mature technologies. The cost-effectiveness of this technology is proven, so it can compete with fossil fuel-based electricity generation in certain environments.

We also provide new insights in the discussion about RPS schemes. Looking at US states, Carley (2009) found a positive effect, whereas Delmas & Montes-Sancho (2011) did not find a significant positive contribution of RPS to RE investments. Thus, we confirm Carley (2009), Bird et al. (2005), and Menz & Vachon (2006) for a sample of countries beyond the USA. On the other hand, we contrast Delmas & Montes-Sancho (2011). Reliable support through codes and standards or obligation schemes supports investments into RE through institutional investors. These policy measures are mostly long term and not easily retractable, so it gives them certainty in their investment calculations.

5.4.6 Robustness checks

We performed a number of robustness checks to verify the stability of our results. First of all we estimated pooled OLS regressions (which have the strongest assumptions with regard to heteroskedasticy and auto-correlation as well as distribution of errors) and calculated the variance inflation factors (VIF). The analysis revealed that only in the model for the solar sector, multi-collinearity might be an issue (mean of VIF 9.17). However, this high value stems from the VIF of our control variables and therefore has no influence on the coefficients of our independent variables of interest (i.e. the policy measures). Second, we estimated random effects estimators (REE) for our models (see Table 8-6). All our models (PCSE, OLS and REE) display consistent results. When using the OLS model, grants and subsidies as well as codes and standards become insignificant. Third, we ran our models, including only the significant variables from our previous analysis. These analysis displayed consistent results throughout all models. Finally, to account for the excessive number of zeros in our sample, we crosschecked the disaggregated results (i.e. individual policy instruments) with aggregated results (categories of policy instruments) for the entire analysis. The models displayed consistent results.

5.5 Conclusions and policy implications

This work contributes to two different streams of the academic literature: Exploring the effectiveness of RE policies as well as observing the role of policies in the decision making process for RE capacity investments through institutional investors. Our research revealed mixed evidence from the wind, solar and biomass sector. We call for technology specific policies, taking into account the actual market conditions and the position in the technology life cycle to design a supportive policy mix. Our results strongly suggest the establishment of a reliable framework with a clear vision and long-term policy objectives regarding the RE capacities to be installed in the future as well as complementary transitions in the energy sector. Ex-post changes to the remuneration of existing projects should be avoided. However, as technological progress continues, the measures taken need to be adjusted, taking the market and technological conditions (i.e. life cycle) into account.

Within this framework, monetary and fiscal and economic incentives are the most relevant policy measures for investors. These directly impact the risk/return profile of RE projects and, thus, their attractiveness. Investors are positive about long-term reliable support mechanisms that cannot be revoked and provide a highly predictable revenue stream. FIT provide more reliable and long-term signal than grants which depend on public budgets. However these funds influence the direct and early project cash flows, which is also seen favourable.

According to our results, market based incentives (such as GHG emission trading systems) can also have strong influence on investments by institutional investors. These measures support the need of investors for a highly reliable environment, best accompanied by a diminished risk exposure. However, for an emission trading system to become an effective anchor for institutional investors, the technology deployed should have reached maturity.

Supportive regulatory measures such as codes and standards (especially RPS) accelerate the diffusion process of RE technologies by further reducing technological and regulatory risk associated with investments in RE projects. Thus we recommend the streamlining and strengthening of legislation and a transparent setting of renewable energy targets.

Our results also provide implications for institutional RE investors who are looking for stable returns unrelated to volatile capital markets. They recognize a regulatory environment which supports their investments or lowers their risks. We suggest them to allocate their funds in countries which have shown to be long-term supporters of the RE markets and have not changed policies abruptly. Higher certainty in the reliability of RE technologies and increasing cost competitiveness make the market increasingly independent from direct support mechanisms. To further accelerate the diffusion the policy focus can move towards a reliable environment for RE.

There are a number of limitations regarding study design and modelling. First, the use of dummy variables for the policy measures does not allow for statements concerning policy implementation, policy design or policy uncertainty (Bergek et al., 2013; Jenner et al., 2013; Lüthi & Wüstenhagen, 2012; Müller et al., 2011). Second, our analysis does not cover the most recent developments in RE deployment and investments (2012 onwards) due to data availability.

Amending our fine grained policy analysis, future studies could look at the influence of these and other policies on general capacity additions among different types of private and institutional investors such as banks, insurances, university endowments, pension funds and family offices as well as including non-institutional investments such as households and utilities. Furthermore, interaction effects between the different policy instruments are worthwhile investigating in a longitudinal research design to discover complementarities and synergies. Geographically our research could be extended to the BRIC countries and less developed countries (LDC) which might alter the results due to an different institutional setting (Friebe et al., 2013, 2014). Finally, it would be interesting to close the link between early stage and later stage financing along the finance value chain for RE technologies, thus analysing the support environment for venture capital and private equity investments in the early and later stages of RE companies which might interact with RE project investments.

Acknowledgements
The authors are grateful for the time and support of Martin Kenney, Donald Patton (University of California, Davis), Alex Coad, and Paul Nightingale (SPRU – University of Sussex). We thank six anonymous reviewers for useful comments

and suggestions. In addition the discussion at the ZEW Energy Conference 2014 helped us in further refining our arguments. We would like to thank the Federal Ministry of Education and Research (BMBF), Germany, for their financial support as part of the research project "Climate Change, Financial Markets and Innovation (CFI)".

6 Conclusions and implications

The aim of this thesis was to explore the diverse set of barriers to low-carbon innovation along the innovation cycle and highlight consequences for finance, which have been addressed exemplarily through three studies. It thereby contributes to the debate surrounding the mobilisation of finance for innovative clean technologies as well as conducive structures (i.e. identifiable organisational, institutional or legal arrangement between actors in the innovation system) and policies.

This chapter presents the broader findings and implications across the literature review and all three empirical studies. Firstly, it summarises barriers to low-carbon innovation and highlights indicative consequences for private finance. Subsequently, the main results are arranged according to structures and policy measures required to address these barriers and effectively and efficiently mobilise private financial resources. Finally, the findings are compared to theory and implications for research and policy makers are drawn.

6.1 Conclusions

6.1.1 Barriers to low-carbon innovation inhibit the financing for companies, projects and infrastructure

The first conclusion following from analyses conducted in this thesis concerns barriers to eco-innovation. Prior research highlights the presence of barriers to the commercialisation and diffusion of eco-innovation due to the characteristics of innovative technologies and environmentally related externalities (Foxon & Pearson, 2008; Iyer et al., 2013; Jaffe et al., 2005). Scholars further regard the removal of institutional and non-institutional barriers to energy efficiency and low-carbon technologies as a crucial element for achieving the transition towards a low-carbon economy (Foxon et al., 2008; Jacobsson & Bergek, 2011; Jefferson, 2008). However the discussion in relevant literature streams has so far focused on market and system failures on an abstract level (Foxon & Pearson, 2007; Jacobsson & Bergek, 2011; Rennings, 2000). A second stream of literature has scrutinised precise barriers in a number of research settings, generating evidence to supporting policy makers (Foxon & Pearson, 2007; Kley et al., 2011; Köhler et al., 2010).

Through the development of a comprehensive innovation cycle framework on the technology level and a categorisation of barriers according to the technological, institutional, economic, financial, political, capability, interactional and transitional context in chapter 2, it became evident that abstract failures highlighted by other researchers can translate into tangible barriers along the innovation cycle. In addition, it became clear that policy makers can choose from a variety of instruments to address these barriers. These are categorised into active technology policy, institutional support, market creation, policy design, fostering interaction and increasing knowledge and learning as well as facilitating transformation. By using the derived categories policy makers can focus on specific barriers at the corresponding stages in the innovation cycle.

By focusing on tangible barriers that relate to finance along the innovation cycle, the chapters 2 and 3 revealed the tremendous potential of connecting public support with private finance in an effective and efficient manner. Thin markets for finance especially present huge opportunities in accelerating the innovation process for clean technologies. These findings are supported by other researchers such as Mowery et al. (2010), Hargadon (2010), Mazzucato (2013) and Kenney & Hargadon (2012). However, chapter 2 also shows that tangible structures and policy instruments remain largely unexplored. Thus, particularly in the case of eco-innovation, more systemic efforts are needed to balance regulation, innovation and complementary financial mechanisms and thereby address lock-in effects, path-dependency and other barriers to commercialisation and diffusion.

6.1.2 Transparent structures which focus on risk and return facilitate private investments into clean technologies along the innovation cycle

The second conclusion based on analysis in this thesis revolves around conducive structures for private investments. Scholars acknowledge that the promotion of generation, commercialisation and diffusion of low-carbon technologies should involve support for research and development (R&D), demonstration and deployment as well as the creation of markets (Foxon et al., 2008; Iyer et al., 2013; van den Bergh, 2013). However specific structures (i.e. organisational, institutional or legal arrangements between actors in the innovation system) to accomplish this task by mobilising private finance have been neglected so far. In order to explore structures in the commercialisation phase, Chapter 3 sheds light on innovation intermediaries as critical links between public and private actors (Kivimaa, 2014; Klerkx & Leeuwis, 2009; van Lente et al., 2003). The chapter highlights the need for an advanced consideration of innovation intermediation. The case of eco-innovation is especially interesting with regard to financing as severe system failures persist.

The analysis shows the value and importance of strengthening the competences and functions of institutional innovation intermediaries as a key component to increase effectiveness and efficiency of public spending. A carefully designed public-private cooperation with critical actors such as financial service providers (e.g. venture capitalists (VCs), banks) reduces information asymmetries along the innovation process for low-carbon innovation and addresses bottlenecks. Thus, innovation intermediation should go beyond mere establishment of markets for technologies, as highlighted by Hoppe & Ozdenoren (2005). More specifically, innovation intermediaries should be empowered to make the technology development process more transparent and to combine private finance with public grants. These measures have proven to be beneficial for SMEs in the context of strategic research partnerships (Audretsch et al., 2002; Chang et al., 2002; Link & Scott, 2010).

Coordinating different sources of capital in the early stages – for example research grants, public-private-partnerships (PPP) and private sources of capital from firms and financiers – to address bottlenecks at each stage of the innovation process could also become a key task for innovation intermediaries. Scholars emphasise the management of funding as a generic function in the commercialisation phase (Howells, 2006; Kivimaa, 2014; Yusuf, 2008). Chapter 3 disentangles this view, showing possible direct financial instruments (e.g. grants and subsidies, start-up support schemes, public procurement) or indirect tools (e.g. socio-economic research and roadmaps) as ways to address financial barriers such as underinvestment in R&D, scalability or capital intensity. Thereby, the establishment of institutional innovation intermediaries represents a tangible structure to facilitate public-private cooperation and to foster markets for early stage technologies and early stage finance.

Moving towards the pre-commercial and commercial stages of the innovation process for clean technologies, structures of public-private-cooperation need to be found that allow partnering and risk sharing between public and private actors. As one example of such a risk sharing mechanism, chapter 4 explores in detail the role of energy service companies (ESCo) and contracting. The case of an innovative energy efficiency technology – LED public street lighting in Germany – allows a full scrutiny of the challenges and opportunities of the mechanism in a stable policy environment. In this regard, the chapter addresses the barriers of technological risks, infrastructure, public acceptance and stakeholder involvement, high upfront investment, cost for deployment and high discount rates for future savings (Bergek et al., 2013; Iyer et al., 2013; Sovacool, 2008; van den Bergh, 2013; Zhang et al., 2012). Energy performance contracting (EPC) is suited to address these barriers, which translate into high transaction costs in the case of LED street lighting.

EPC could be a public-private arrangement to use private sector competencies when procuring innovative products in a public context. Hence this chapter extends previous work that emphasises contracting as a commercially viable way to increase the diffusion of innovative technologies in the manufacturing industry (Hypko et al., 2010a, 2010b). Additionally it provides the opportunity to serve as risk management in a municipal context (Jackson, 2010).

Overall this thesis highlights public-private structures that allow for risk sharing, increasing transparency and the creation of markets to accelerate the commercialisation process. These mechanisms do not compromise on the selection mechanism of early technology markets. They represent conducive market-based vehicles to mobilise private finance for the transition towards low-carbon innovation.

6.1.3 Science, technology and innovation (STI) policies and regulation are needed to spur private investments into clean technologies

The third conclusion that can be drawn from research conducted in this thesis is the need for an analysis of policy measures to overcome specific barriers to low-carbon innovation. This corresponds to for the requirement of additional studies focussing on the effectiveness and efficient policy instruments to stimulate private investments into innovative clean technologies such as companies, projects and infrastructure. These policy responses should be tailored to the actual phase in the innovation cycle (see chapter 2 and Iyer et al., 2013; Jaffe et al., 2002; Newell et al., 2006).

In the early stages of the innovation cycle, STI policies affect the innovation process for clean technologies (Kivimaa & Mickwitz, 2006; Mickwitz et al., 2008). Typical STI policies are research programs and partnerships as well as regulatory instruments that create a suitable institutional environment. However, these policies need to be coordinated or synchronised in order to allow for successful market development. The linkages between STI policy and regulations can be addressed on a sectoral level, focussing on individual technologies and their potential application. Innovation intermediaries at the intersection between public, policy and private actors facilitate coordination and reduce reflexivity failures (Kivimaa, 2014; Klerkx & Leeuwis, 2009; van Lente et al., 2003; Weber & Rohracher, 2012). In other words, intermediaries reduce information asymmetries between the needs of the private sector and public policy making.

By looking at the case of LED street lighting in Germany, a shift in public involvement could be explored throughout chapter 4. More specifically, the case applies a longitudinal research design to show the transition from research subsidies in the early stages towards demonstration projects and deployment subsidies in the later stages. The public market for street lighting proves relevant, thus public procurement as a mission-oriented innovation policy is adequate (Edquist & Zabala-Iturriagagoitia, 2012; Uyarra et al., 2014). However public budgets are limited, hence new solutions which simultaneously support the diffusion of LED lighting and mobilise private investments need to be found. Designing the conditions for PPP is therefore a crucial task for policy makers, which requires competencies and institutional support in the form of an energy service market. This chapter shows that a combination of public-procurement (Edquist & Zabala-Iturriagagoitia, 2012), performance based regulation (Popp, Hafner, et al., 2011) and the development of a PPP-structure such as EPC accelerates the commercialisation of energy efficiency innovation.

During the later stages of the innovation cycle a number of barriers inhibit the deployment of mature technologies, such as renewable energy (RE). These include high upfront investment, capital intensity, regulatory uncertainty and risks, infrastructure and grid connection, long timescales as well as slow capital stock turnover and corresponding long payback periods (Gallagher et al., 2006; Iyer et al., 2013; Sandén & Azar, 2005; Zhang et al., 2012). Policy makers are confronted with a variety of policy options to support the deployment of clean technologies. These range from establishing a system which puts a price on carbon and enables market actors over trading the permits towards more direct instruments, especially subsidies, regulation, taxes or PPP-investments (Foxon et al., 2008; Mathews et al., 2010; Mitchell et al., 2006; Quirion, 2010). By analysing the effectiveness of different policy instruments to attract private institutional investors into RE projects, chapter 5 provides a relevant contribution to this debate (Bergek et al., 2013; Friebe et al., 2014; Wüstenhagen & Menichetti, 2012). Technology-specific policies need to be established, taking into account the actual market conditions and the position in the technology life cycle. Our results strongly suggest the establishment of a reliable framework with a clear vision and long-term policy objectives regarding RE capacities to be installed in the future as well as complementary transitions in the energy sector.

Within this framework, fiscal and economic incentives are the most relevant policy measures for investors. Feed-in tariffs (FIT) provide a more reliable and long-term signal than grants which often depend on public budgets. FITs therefore attract institutional investors into less mature technologies such as solar photovoltaic However, loans or loan guarantee programs do not exhibit a positive influence on solar investments. Beyond, market-based incentives such as greenhouse gas (GHG) emission trading systems also show a strong influence on investments by institutional investors. These measures support the need of investors for a highly reliable environment, best accompanied by a diminished risk exposure. The findings fit only for technologies that have reached a certain level of maturity, such as wind energy and biomass, ruling out technological difficulties. However, market-based incentives do not prove effective in the case of solar technologies. In addition, supportive regulatory measures such as renewable portfolio standards (RPS) accelerate the diffusion process of RE technologies by further reducing technological and regulatory risks.

This thesis provides the opportunity to compare policy instruments used in early and later stages of the innovation cycle across a range of technologies. On the one hand, in the transition towards commercialisation and early commercialisation stages, technological risks and uncertainty as well as competence problems or information asymmetries prevail, whereas during the later stages, risks related to the institutional, market and political environment dominate. On the other hand, energy efficiency and RE technologies exhibit different characteristics with regard to finance. The former requires a large number of small amounts combined with a high technological competence tailored to the setting which prevents effective scaling (Schleich, 2009; Sorrell et al., 2004). Energy efficiency investments depend on a range of factors including volatile energy prices. Yet, they exhibit less dependency on political regulations. The latter permits the investor to scale a project easily, so that large institutional investors could provide the funds to a potentially huge market. RE investments largely depend on the competitiveness with non-renewable energy sources, thus the policy environment shapes their commercial viability significantly (Friebe et al., 2014; Lüthi & Prässler, 2011; Wüstenhagen & Menichetti, 2012). The common perspective centres on the perception of risk and returns by private actors. Policy makers adopting this perspective could adjust their interventions in order to attract private finance according to the development stage of the targeted clean technology. Therefore synchronising STI policy and regulation and selecting the appropriate intervention such as public procurement, other forms of public-private cooperation or incentives for the private sector is needed.

6.2 Implications for research and policy makers

6.2.1 Implications for research

The analyses presented in this thesis have implications for research on low-carbon innovation, the corresponding finance mechanisms and the policy environment. Overarching contributions to a number of related literature streams in innovation studies are presented in the following paragraphs. More specifically, the results contribute to the debate about how to address barriers to eco-innovation while mobilising private finance through tangible structures and policy instruments.

At first, this thesis shows that abstract failures and big crises in the background could only be addressed on a very concrete level. Reconciling literature streams provides a holistic understanding of the public-private interplay in the innovation-policy-finance nexus for clean technologies. Chapter 2 complements earlier work that focuses on the abstract failures to innovation on a systemic and technology level (Edquist, 2011; Jacobsson & Bergek, 2011; Klein Woolthuis et al., 2005; Weber & Rohracher, 2012) as well as on concrete technology oriented research (Foxon & Pearson, 2007; Kley et al., 2011; Köhler et al., 2010). More specifically this chapter shows that barriers to low-carbon innovation have different effects on the innovation process along the technology life cycle, which means that policy responses need to be tailored to fit the actual technology and market conditions. Connecting the abstract failures and tangible barriers has been missing throughout the literature (Dewald & Truffer, 2011; Newell et al., 2006; van den Bergh, 2013). Vis-à-vis, selected barriers have an influence on the ability of private financiers to supply resources for innovative companies, infrastructure and projects. Thus addressing these barriers and maximising private investments requires an understanding of the logic behind financiers' perception of the innovation process for clean technologies.

The structural integration of a perspective on finance in the early stages of the innovation process is crucial for effective and efficient policy support. This perspective includes a risk and return calculus, a focus on commercialisation and possible influences of STI and the regulatory environment on the financing eco-system of novel technologies. Thereby, chapter 3 extends existing research on innovation intermediaries such as Kivimaa (2014), Klerkx & Leeuwis (2009), Stewart & Hyysalo (2008) and Howells (2006). Institutional intermediaries with their 'financial mobilisation functions' are able to significantly accelerate the commercialisation and diffusion of eco-innovation. Financial mobilisation functions that comprise direct instruments and cooperation with financiers may catalyse the functioning of other innovation policy instruments, and vice versa.

From the analysis in chapter 4, it follows that a combination of structural elements such as public-private-partnerships and mission-oriented procurement policies are able to successfully address barriers in the commercialisation and diffusion stages. In view of the existing literature, these measures translate into reduced transaction costs (Edquist & Zabala-Iturriagagoitia, 2012; Schleich, 2009; Sorrell et al., 2004; Uyarra et al., 2014). Theoretically, this chapter combines innovation studies literature (especially the role of public procurement) and literature on ESCo focusing on transaction costs. This perspective is fruitful to understand the underlying barriers to commercialisation and diffusion of eco-innovations. Possible solutions include modes of governance to introduce and apply energy efficiency innovation (Delmas, 1999; Hartmann et al., 2014; Roehrich et al., 2014). These modes of governance are integrated into a theoretical model for energy service commercialisation and diffusion. The transaction cost perspective is able to explain both the slow diffusion of innovative energy efficiency technologies as well as strategies and structures to overcome corresponding barriers.

Technology and context specific policy measures are not only important for early stage innovations but also for mature clean tech innovations such as RE, as chapter 5 shows. Depending on the market conditions and the phase in the technology life cycle, it incorporates a stable institutional environment, fiscal and financial incentives such as feed-in tariffs and market-based instruments such as an emission trading system. This thesis thereby confirms prior literature that emphasises the perspective of institutional investors as relevant to spur the deployment of RE technologies (Bergek et al., 2013; Bürer & Wüstenhagen, 2009; Wüstenhagen & Menichetti, 2012) and provides a differentiated view on policy instruments consolidating earlier work on the effectiveness of policy instruments in large longitudinal settings (Aguirre & Ibikunle, 2014; Johnstone et al., 2010; Marques & Fuinhas, 2012a).

Overall this thesis has implications for the transformation phase of innovation systems. In the current installation period of the 'green economy' with ongoing experimentation, policy makers need to redirect capital flows back into production (Perez, 2009, 2013). Chapters 3–5 show possibilities to connect public policies and private finance by establishing tangible structures to overcome barriers to low-carbon innovation. The research findings reported here provide a possibility to integrate the wider finance perspective – such as the focus on risk, uncertainty, and returns or gains – into the innovation policy literature (Foxon et al., 2008; Mazzucato, 2013). Thereby, policy makers are able to create markets for eco-innovations more effectively and efficiently.

This thesis is also an attempt to start integrating financial and innovation studies perspectives. Prior research suggested that this connection has been neglected in the past (Dosi, 1990; O'Sullivan, 2006; Perez, 2002). Therefore, tangible cases along the innovation cycle are needed to explore this connection more holistically and thus contribute to a broader understanding of the different logics behind finance and innovation. This thesis enriches the ongoing debate by including aspects like risk and return and the fundamentally uncertain outcomes of innovation processes into the concept of failures and barriers in IS research (Bergek et al., 2008, 2013; Hekkert & Negro, 2009).

6.2.2 Implications for policy makers

This thesis has several implications for policy makers. First, when addressing barriers to low-carbon innovation, policy makers need to pay attention to the actual stage in the innovation cycle and the corresponding market to tailor their policy intervention while maximising private investments. Considering the interaction between institutional, economic and transformational barriers while accounting for the political barriers such as coordination and reflexivity failures has implications for the financing of clean technology firms, projects and infrastructure. Economic, technological and knowledge barriers translate into private under-investments in clean R&D in the early stages. Changes to support mechanisms and missing complementary assets (such as infrastructure) significantly impact the ability to obtain private finance during commercialisation. Regulatory changes and power of incumbents applying fossil-fuel based technologies hinder private financiers from investing even in mature technologies due to an uncertain market outlook.

Second, in order to mobilise private investments, policy makers need to reflect on the implication for the risk and return calculus of financiers before designing a market intervention and deciding which kinds of innovation to support. The ability of private financiers to allocate resources for clean technologies should not be hampered by policy intervention. Thus engaging in an active collaboration with private financiers is crucial to provide a mutual understanding of the innovation process for clean technologies, its potential risks and future market opportunities, thereby reducing information asymmetries. This collaboration should start at the verge towards the commercialisation of a technology by designing research partnerships that tackle these financial questions together with technology development. For example, research projects that include socio-economic analysis of technology-based innovations could be a useful structure to analyse public and private risks and rewards to support policy makers in their decision making. An engagement by institutional innovation intermediaries proves beneficial to accomplish this task. Relevance, possibilities and advisability of including questions related to finance in the early stages of the innovation cycle need to become central to STI policy.

Third, investments into green and sustainable companies, projects and infrastructure need to be encouraged in order to address the broader implications of climate change. Therefore, scholars suggest the matching of innovation potential and demand (Mazzucato & Perez, 2014; Perez, 2013). In this respect, this thesis highlights the positive role of well-designed PPPs as suitable structures that distribute the risks and rewards effectively to encourage innovation and diffusion of clean technologies. Channelling the abundant financial resources into companies, projects and infrastructure based on innovative clean technologies requires new forms of cooperation and capabilities both on the public and the private sides. Partnering of public and private actors via energy service contracting emerges as a suitable mechanism to accelerate the commercialisation and diffusion during market phases of clean technologies. Thus government bodies can act as drivers for innovation without spending large amounts of public resources for market development. These activities could be embedded in an overall lead market design along the innovation cycle.

Fourth, to complement such activities, a strong government supporting mission oriented R&D and innovation is needed. This is especially valid since financial pressures remain high after the financial and budgetary crisis (Lazonick & Mazzucato, 2013). Hence policy makers should start with a clear strategic vision in the various clean technology sub-sectors and encourage private investments, synchronising STI policy and corresponding regulation, and providing direct fiscal and financial incentives as well as market based incentives to accompany a technology stream or sector from the early stages towards maturity. To maximise private investments, regulatory changes need to be adjusted according to technological improvements. Embedding these changes in a transparent consultation process involving policy makers and private actors provides the necessary reliability vis-à-vis financiers.

Fifth, this thesis provides concrete evidence that in order to foster green innovation in the real economy, medium and long-term orientated finance, monitoring and controlling as well as services in the public sector are needed (Mazzucato, 2013). On the one hand, policy makers need to evaluate how the barriers along the innovation cycle of clean technologies are perceived by financiers and which incentives or regulation targeting both real economy and financial markets could foster their engagement. On the other hand, financiers should sharpen their competencies with regard to concrete technologies, business models and policy initiatives to develop new methods of financing innovative clean technologies. Policy makers could assist this process of by making the technology development process more transparent, i.e. identifying future finance needs and thus transforming uncertainty into calculable risk and returns (Flotow & Schiereck, 2013).

The overarching questions addressed in this thesis relate to the debate about the future of capitalism, especially the role of public policy and private financial markets (Mazzucato & Perez, 2014). Innovation is a combination of new technologies and accumulation of finance that happens in a cyclical form. Perez (2013) argues that the market for clean technologies is at a turning point. Thus, a risk averse private sector with volatile private investment needs a strong government with clear support and mission oriented financing and investment. This thesis contributes to the debate by highlighting tangible structures and policy instruments in the commercialisation and diffusion stages for clean technology innovation.

By facilitating green growth predominantly financed by the private sector, policy makers can address the current climate crisis, financial crises and related growth problems effectively. Consequently the separation of financial markets and the real economy needs to be overcome. This could be done by increasing public-private interaction, transparency and the creation of markets with a stronger focus on the financing environment to mobilise private investments.

6.3 Limitations

Every research has its limitations and this thesis acknowledges its own. Above all, the author is aware of the fact that the debate about limits to growth on a finite planet is ongoing for over 40 years (Meadows, Goldsmith, & Meadows, 1972; Meadows, Randers, & Meadows, 2004). In order to mitigate climate change via technological change and decouple economic growth and development from the use of finite resources and fossil fuels (i.e. the diffusion of low-carbon and energy-efficient technologies), more fundamental changes in patterns of lifestyles and consumption are required (Foxon et al., 2008). These aspects are clearly beyond the scope of this thesis. In addition, climate change mitigation and adaptation has complex social implications e.g. regarding inequality within the developed world and between the global north and the global south (Lazonick & Mazzucato, 2013; Perez, 2004). These questions are equally beyond the scope of the research reported here.

This thesis exemplarily investigates structures and policies to mobilise private finance for clean technology innovation. The author acknowledges that the chosen cases and methods do not provide a comprehensive treatment of the occurring barriers and policy responses although the relevant examples are investigated both from a research and policy perspective.

The literature review conceptually provides the possibility to examine barriers to low-carbon innovation from a range of angles in innovation studies, including technology and transition. Hence the focus on finance represents only one avenue for future research. Moreover the narrative character of the review does not allow making inferences from connections of authors and literature streams which would require a bibliometric analysis.

Empirical limitations stem from the qualitative and explorative research designs in chapter 3 and 4 that focus on the institutional setting in Germany.

Hence, the findings exhibit limited transferability to other countries. Chapter 3 covers a variety of technologies thus allowing comparability. However, intermediation as a function in innovation systems is a context-dependent phenomenon which requires further investigations of similar organisations in other empirical settings to validate and extend the results. Chapter 4 focuses on one particular technology that only assents to conclusions for similar energy efficiency applications. The role of EPC, which has been investigated, equally depends on the market structure and boundary conditions of the energy system which limits transferability. Finally the analysis in chapter 5 provides a quantitative empirical investigation of the influence of policies on subsequent investment. The research approach taken is limited with regard to its timeframe as it does not cover the most recent developments in RE investments and RE investments in all countries due to data availability. Methodologically the dummy variables used for the policy indicators do not permit to evaluate differences in policy design. In addition, the research was limited to three relevant technologies in the RE sector.

6.4 Future research

Based on the limitations of the individual empirical articles and the literature review, this thesis opens up a research agenda on financing low-carbon innovation by exploring other structures and policy mechanisms to mobilise private finance. Non-financial barriers play out in the finance environment, which necessitates further exploration of the interaction of policy measures to spur eco-innovation and its consequences for financial markets. Thus additional research on other structures to finance low-carbon innovation and interaction with STI policy and regulatory measures could be fruitful.

First, other structures to apply innovative technologies, and increase transparency could be explored such as additional forms of servitization and business models for innovative clean technologies that allow risk sharing and the use of private sector competencies (Hannon et al., 2013; Kley et al., 2011; Steinberger et al., 2009). Amending the current research in chapter 4, this could be done by exploring structures in research settings beyond Germany or in a quantitative survey design. Additionally, the innovation intermediaries literature stream used in chapter 3 (Kivimaa, 2014; Klerkx & Leeuwis, 2009; van Lente et al., 2003) has proven particularly useful to study the interaction of finance mechanisms and clean technology development and commercialisation. Complementing the current research with a survey would be a promising direct line of research. Beyond, further research on the peculiarities of eco-innovation with regard to innovation intermediaries would be valuable.

Second, on the national level, PPP fund structures that provide VC and private equity as well as project finance could be explored to complement risk averse private capital in the early and later stages (Mathews et al., 2010). In addition, crowdfunding platforms as a structure for micro-finance of companies or projects that develop and apply innovative clean technologies would be interesting for further research. On an international level, research on structures to channel public and private money into clean technologies, such as the Green Climate Fund (GCF) should be evaluated with regard to their innovation effect to foster long-term green growth (Flotow & Schiereck, 2013; Friebe et al., 2014; Mathews et al., 2010).

Third, research on a policy mix to foster long-term low-carbon innovation systems including transformational elements proves useful in order to give recommendations on how the ongoing clean technology revolution could be governed (Flanagan, Uyarra, & Laranja, 2011; Guerzoni & Raiteri, 2014; Kern & Smith, 2008). Furthermore the link between policy instruments, regulation of financial markets and private finance mechanisms could be analysed regarding complementarity and synergies (Migendt et al., 2014).

This thesis provides a contribution to the ongoing debate on effective and efficient structures and policy mechanisms for the much needed transformation towards a 'green economy'. Future research may build on these findings, challenge and broaden them, engaging in an active academic and policy debate on how to mitigate climate change and foster economic development at the same time.

7 References

Acemoglu, D., Aghion, P., Bursztyn, L., & Hemous, D. (2012). The Environment and Directed Technical Change. *American Economic Review, 102*(1), 131–166. http://doi.org/10.1257/aer.102.1.131

Aguirre, M., & Ibikunle, G. (2014). Determinants of renewable energy growth: A global sample analysis. *Energy Policy, 69*, 374–384. http://doi.org/10.1016/j.enpol.2014.02.036

Akerlof, G. A. (1970). The Market for "Lemons": Quality Uncertainty and the Market Mechanism. *The Quarterly Journal of Economics, 84*(3), 488–500. http://doi.org/10.2307/1879431

Angrist, J. D., & Pischke, J.-S. (2008). *Mostly Harmless Econometrics: An Empiricist's Companion*. Princeton University Press.

Arabatzis, G., & Myronidis, D. (2011). Contribution of SHP Stations to the development of an area and their social acceptance. *Renewable and Sustainable Energy Reviews, 15*(8), 3909–3917. http://doi.org/10.1016/j.rser.2011.07.026

Arent, D. J., Wise, A., & Gelman, R. (2011). The status and prospects of renewable energy for combating global warming. *Energy Economics, 33*(4), 584–593. http://doi.org/10.1016/j.eneco.2010.11.003

Arrow, K. J. (1962). Economic Welfare and The Allocation of Resources for Invention. In R. R. Nelson (Ed.), *The Rate and Direction of Inventive Activity, Princeton University Press and NBER*.

Audretsch, D. B., & Lehmann, E. E. (2004). Financing High-Tech Growth: The Role of Banks and Venture Capitalists. *SSRN eLibrary*. Retrieved from http://papers.ssrn.com/sol3/papers.cfm?abstract_id=630982

Audretsch, D. B., Link, A. N., & Scott, J. T. (2002). Public/private technology partnerships: evaluating SBIR-supported research. *Research Policy, 31*(1), 145–158. http://doi.org/16/S0048-7333(00)00158-X

Auerswald, P. E., & Branscomb, L. M. (2003). Valleys of death and Darwinian seas: Financing the invention to innovation transition in the United States. *The Journal of Technology Transfer, 28*(3), 227–239. http://doi.org/10.1023/A:1024980525678

Babl, C., Schiereck, D., & Flotow, P. von. (2012). Clean technologies in German economic literature: a bibliometric analysis. *Review of Managerial Science, 8*(1), 63–88. http://doi.org/10.1007/s11846-012-0095-8

Backlund, S., & Eidenskog, M. (2013). Energy service collaborations—it is a question of trust. *Energy Efficiency, 6*(3), 511–521. http://doi.org/10.1007/s12053-012-9189-z

Baker, E., Chon, H., & Keisler, J. (2009). Advanced solar R&D: Combining economic analysis with expert elicitations to inform climate policy. *Energy Economics, 31, Supplement 1,* S37–S49. http://doi.org/10.1016/j.eneco.2007.10.008

Bañales-López, S., & Norberg-Bohm, V. (2002). Public policy for energy technology innovation: A historical analysis of fluidized bed combustion development in the USA. *Energy Policy, 30*(13), 1173–1180. http://doi.org/10.1016/S0301-4215(02)00013-7

Barradale, M. J. (2010). Impact of public policy uncertainty on renewable energy investment: Wind power and the production tax credit. *Energy Policy, 38*(12), 7698–7709. http://doi.org/10.1016/j.enpol.2010.08.021

Barreto, L., & Kemp, R. (2008). Inclusion of technology diffusion in energy-systems models: some gaps and needs. *Journal of Cleaner Production, 16*(1, Supplement 1), S95–S101. http://doi.org/10.1016/j.jclepro.2007.10.008

Bauman, Y., Lee, M., & Seeley, K. (2008). Does Technological Innovation Really Reduce Marginal Abatement Costs? Some Theory, Algebraic Evidence, and Policy Implications. *Environmental and Resource Economics, 40*(4), 507–527. http://doi.org/10.1007/s10640-007-9167-7

Beerepoot, M., & Beerepoot, N. (2007). Government regulation as an impetus for innovation: Evidence from energy performance regulation in the Dutch residential building sector. *Energy Policy, 35*(10), 4812–4825. http://doi.org/10.1016/j.enpol.2007.04.015

Beise, M., & Rennings, K. (2004). Lead markets and regulation: a framework for analyzing the international diffusion of environmental innovations. *Ecological Economics, 52,* 5–17. http://doi.org/10.1016/j.ecolecon.2004.06.007

Bennett, J., & Iossa, E. (2006). Delegation of Contracting in the Private Provision of Public Services. *Review of Industrial Organization, 29*(1–2), 75–92. http://doi.org/10.1007/s11151-006-9110-z

Bennich, P., Schoenen, B., Scholand, M., & Borg, N. (2014). *Test Report – Clear, Non-Directional LED Lamps.* Retrieved from http://www.eceee.org/ecodesign/products/domestic_lighting/Report_on_Testing_ClearLED_lamps.v5.5a.pdf

Bergek, A., & Jacobsson, S. (2010). Are tradable green certificates a cost-efficient policy driving technical change or a rent-generating machine? Lessons from Sweden 2003–2008. *Energy Policy, 38*(3), 1255–1271. http://doi.org/10.1016/j.enpol.2009.11.001

Bergek, A., Jacobsson, S., Carlsson, B., Lindmark, S., & Rickne, A. (2008). Analyzing the functional dynamics of technological innovation systems: A scheme of analysis. *Research Policy, 37*(3), 407–429. http://doi.org/10.1016/j.respol.2007.12.003

Bergek, A., Mignon, I., & Sundberg, G. (2013). Who invests in renewable electricity production? Empirical evidence and suggestions for further research. *Energy Policy*, *56*, 568–581. http://doi.org/10.1016/j.enpol.2013.01.038

Bergek, A., & Onufrey, K. (2013). Is one path enough? Multiple paths and path interaction as an extension of path dependency theory. *Industrial and Corporate Change*, dtt040. http://doi.org/10.1093/icc/dtt040

Bertoldi, P., Boza-Kiss, B., Panel, S., & Labanca, N. (2014). *ESCO Market Report 2013,*. Ispra, Italy.

Betsill, M. M., & Bulkeley, H. (2006). Cities and the Multilevel Governance of Global Climate Change. *Global Governance: A Review of Multilateralism and International Organizations*, *12*(2), 141–159. http://doi.org/10.5555/ggov. 2006.12.2.141

Bird, L. A., Bolinger, M., Gagliano, T., Wiser, R., Brown, M., & Parsons, B. (2005). Policies and market factors driving wind power development in the United States. *Energy Policy*, *33*(11), 1397–1407. http://doi.org/10.1016/j. enpol.2003.12.018

Bird, L. A., Holt, E., & Levenstein Carroll, G. (2008). Implications of carbon cap-and-trade for US voluntary renewable energy markets. *Energy Policy*, *36*(6), 2063–2073. http://doi.org/10.1016/j.enpol.2008.02.009

Blanford, G. J. (2009). R&D investment strategy for climate change. *Energy Economics*, *31*, *Supplement 1*, S27–S36. http://doi.org/10.1016/j.eneco.2008.03.010

Bleda, M., & del Río, P. (2013). The market failure and the systemic failure rationales in technological innovation systems. *Research Policy*, *42*(5), 1039–1052. http://doi.org/10.1016/j.respol.2013.02.008

Bleyl, J. W., Adilipour, N., Bareit, M., Kempen, G., Cho, S.-H., Vanstraelen, L., & Knowledgecenter, F. (2013). ESCo market development: A role for Facilitators to play.

Blyth, W., Bradley, R., Bunn, D., Clarke, C., Wilson, T., & Yang, M. (2007). Investment risks under uncertain climate change policy. *Energy Policy*, *35*(11), 5766–5773. http://doi.org/10.1016/j.enpol.2007.05.030

BMBF. (2014). LED Leitmarktinitiative. Retrieved November 5, 2014, from http://www.photonikforschung.de/forschungsfelder/leuchtdioden-led/led-leitmarktinitiative/

BNEF. (2013). *Bloomberg New Energy Finance (BNEF)*. Bloomberg. Retrieved from http://www.bnef.com

Böhringer, C., Mennel, T. P., & Rutherford, T. F. (2009). Technological change and uncertainty in environmental economics. *Energy Economics, 31, Supplement 1*, S1–S3. http://doi.org/10.1016/j.eneco.2009.05.006

Boje, D. M. (2001). *Narrative Methods for Organizational and Communication Research*. SAGE.

Bolkesjø, T. F., Eltvig, P. T., & Nygaard, E. (2014). An Econometric Analysis of Support Scheme Effects on Renewable Energy Investments in Europe. *Energy Procedia, 58,* 2–8. http://doi.org/10.1016/j.egypro.2014.10.401

Boon, W. P. C., Moors, E. H. M., Kuhlmann, S., & Smits, R. E. H. M. (2008). Demand articulation in intermediary organisations: The case of orphan drugs in the Netherlands. *Technological Forecasting and Social Change, 75*(5), 644–671. http://doi.org/10.1016/j.techfore.2007.03.001

Boon, W. P. C., Moors, E. H. M., Kuhlmann, S., & Smits, R. E. H. M. (2011). Demand articulation in emerging technologies: Intermediary user organisations as co-producers? *Research Policy, 40*(2), 242–252. http://doi.org/10.1016/j.respol.2010.09.006

Borrás, S., & Edquist, C. (2013). The choice of innovation policy instruments. *Technological Forecasting and Social Change, 80*(8), 1513–1522. http://doi.org/10.1016/j.techfore.2013.03.002

Bosetti, V., & Tavoni, M. (2009). Uncertain R&D, backstop technology and GHGs stabilization. *Energy Economics, 31, Supplement 1,* S18–S26. http://doi.org/10.1016/j.eneco.2008.03.002

Brown, J., & Hendry, C. (2009). Public demonstration projects and field trials: Accelerating commercialisation of sustainable technology in solar photovoltaics. *Energy Policy, 37*(7), 2560–2573. http://doi.org/10.1016/j.enpol.2009.01.040

Brown, M. A. (2001). Market failures and barriers as a basis for clean energy policies. *Energy Policy, 29*(14), 1197–1207. http://doi.org/10.1016/S0301-4215(01)00067-2

Bürer, M. J., & Wüstenhagen, R. (2009). Which renewable energy policy is a venture capitalist's best friend? Empirical evidence from a survey of international cleantech investors. *Energy Policy, 37,* 4997–5006. http://doi.org/10.1016/j.enpol.2009.06.071

Butler, L., & Neuhoff, K. (2008). Comparison of feed-in tariff, quota and auction mechanisms to support wind power development. *Renewable Energy, 33,* 1854–1867.

Cagno, E., & Trianni, A. (2014). Evaluating the barriers to specific industrial energy efficiency measures: an exploratory study in small and medium-sized enterprises. *Journal of Cleaner Production, 82,* 70–83. http://doi.org/10.1016/j.jclepro.2014.06.057

Cameron, A. C., & Trivedi, P. K. (2009). *Microeconometrics using stata* (Vol. 5). Stata Press College Station, TX. Retrieved from http://stata.biz/news/statanews.23.4.pdf

Cansino, J. M., Pablo-Romero, M. del P., Román, R., & Yñiguez, R. (2010). Tax incentives to promote green electricity: An overview of EU-27 countries. *Energy Policy*, *38*(10), 6000–6008. http://doi.org/10.1016/j.enpol.2010.05.055

Cantono, S., & Silverberg, G. (2009). A percolation model of eco-innovation diffusion: The relationship between diffusion, learning economies and subsidies. *Technological Forecasting & Social Change*, *76*, 487–496. http://doi.org/10.1016/j.techfore.2008.04.010

Cárdenas-Rodríguez, M., Johnstone, N., Haščič, I., Silva, J., & Ferey, A. (2013). Inducing private finance for renewable energy projects: Evidence from microdata. *OECD Environment Working Papers*.

Carley, S. (2009). State renewable energy electricity policies: An empirical evaluation of effectiveness. *Energy Policy*, *37*(8), 3071–3081. http://doi.org/10.1016/j.enpol.2009.03.062

Carlsson, B., & Stankiewicz, R. (1991). On the nature, function and composition of technological systems. *Journal of Evolutionary Economics*, *1*(2), 93–118. http://doi.org/10.1007/BF01224915

Chadha, A. (2011). Overcoming Competence Lock-In for the Development of Radical Eco-Innovations: The Case of Biopolymer Technology. *Industry and Innovation*, *18*(3), 335–350. http://doi.org/10.1080/13662716.2011.561032

Chang, C., Shipp, S., & Wang, A. (2002). The Advanced Technology Program: A public-private partnership for early stage technology development. *Venture Capital*, *4*(4), 363–370. http://doi.org/10.1080/1369106022000028262

Chassot, S., Hampl, N., & Wüstenhagen, R. (2014). When energy policy meets free-market capitalists: The moderating influence of worldviews on risk perception and renewable energy investment decisions. *Energy Research & Social Science*, *3*, 143–151. http://doi.org/10.1016/j.erss.2014.07.013

Coenen, L., & Diaz Lopez, F. J. (2010). Comparing systems approaches to innovation and technological change for sustainable and competitive economies: an explorative study into conceptual commonalities, differences and complementarities. *Journal of Cleaner Production*, *18*(12), 1149–1160. http://doi.org/10.1016/j.jclepro.2010.04.003

Cooper, R. S. (2003). Purpose and performance of the small business innovation Research (SBIR) program. *Small Business Economics*, *20*(2), 137–151. http://doi.org/10.1023/A:1022212015154

Couture, T., & Gagnon, Y. (2010). An analysis of feed-in tariff remuneration models: Implications for renewable energy investment. *Energy Policy*, *38*(2), 955–965. http://doi.org/10.1016/j.enpol.2009.10.047

Creswell, J. W., & Miller, D. L. (2000). Determining validity in qualitative inquiry. *Theory into Practice*, *39*(3), 124–130. http://doi.org/10.1207/s15430421tip3903_2

Czarnitzki, D., Hanel, P., & Rosa, J. M. (2011). Evaluating the impact of R&D tax credits on innovation: A microeconometric study on Canadian firms. *Research Policy*, *40*(2), 217–229. http://doi.org/10.1016/j.respol.2010.09.017

Dahlstrand, Å. L., & Cetindamar, D. (2000). The dynamics of innovation financing in Sweden. *Venture Capital*, *2*(3), 203–221. http://doi.org/10.1080/13691060050135082

Da Rin, M., Nicodano, G., & Sembenelli, A. (2006). Public policy and the creation of active venture capital markets. *Journal of Public Economics*, *90*(8–9), 1699–1723. http://doi.org/10.1016/j.jpubeco.2005.09.013

David, R. J., & Han, S.-K. (2004). A systematic assessment of the empirical support for transaction cost economics. *Strategic Management Journal*, *25*(1), 39–58. http://doi.org/10.1002/smj.359

De Jager, D., Klessmann, C., Stricker, E., Winkel, T., de Visser, E., Koper, M., … Bouillé, A. (2011). *Financing Renewable Energy in the European Energy Market* (No. PECPNL084659) (p. 264). Ecofys, Fraunhofer ISI, TU Vienna EEG, Ernst & Young.

De Jager, D., Rathmann, M., Klessmann, C., Coenraads, R., Colamonico, C., & Buttazzoni, M. (2008). Policy instrument design to reduce financing costs in renewable energy technology projects. *Ecofys, by Order of the IEA Implementing Agreement on Renewable Energy Technology Deployment (RETD), Utrecht, Ther Netherlands*. Retrieved from http://iea-retd.org/archives/publications/policy-instrument-design

Delmas, M. A. (1999). Exposing strategic assets to create new competencies: the case of technological acquisition in the waste management industry in Europe and North America. *Industrial and Corporate Change*, *8*(4), 635–671. http://doi.org/10.1093/icc/8.4.635

Delmas, M. A., & Montes-Sancho, M. J. (2011). U.S. state policies for renewable energy: Context and effectiveness. *Energy Policy*, *39*(5), 2273–2288. http://doi.org/10.1016/j.enpol.2011.01.034

del Río, P., & Bleda, M. (2012). Comparing the innovation effects of support schemes for renewable electricity technologies: A function of innovation approach. *Energy Policy*, *50*, 272–282. http://doi.org/10.1016/j.enpol.2012.07.014

dena. (2012). *Abschlussbericht „Umfrage: Energieeffiziente Straßenbeleuchtung"*. Retrieved from http://www.stromeffizienz.de/fileadmin/user_upload/dienstleister/beleuchtung/dateien/Abschlussbericht-Umfrage-Strassenbeleuchtung-kurz.pdf

dena. (2015). Energy-Efficient Municipalities. Retrieved from http://www.dena.de/en/projects/building/energy-efficient-municipalities.html

Dewald, U., & Truffer, B. (2011). Market Formation in Technological Innovation Systems—Diffusion of Photovoltaic Applications in Germany. *Industry and Innovation, 18*(3), 285–300. http://doi.org/10.1080/13662716.2011.561028

Difu. (2014). *Finanzierung kommunaler Klimaschutzmassnahmen.* Berlin: Deutsches Institut fuer Urbanistik.

Dinica, V. (2006). Support systems for the diffusion of renewable energy technologies—an investor perspective. *Energy Policy, 34*(4), 461–480. http://doi.org/10.1016/j.enpol.2004.06.014

Dinica, V. (2008). Initiating a sustained diffusion of wind power: The role of public–private partnerships in Spain. *Energy Policy, 36*(9), 3562–3571. http://doi.org/10.1016/j.enpol.2008.06.008

Dosi, G. (1990). Finance, innovation and industrial change. *Journal of Economic Behavior & Organization, 13*(3), 299–319. http://doi.org/10.1016/0167-2681(90)90003-V

Dougherty, D. (2002). Grounded theory research methods. In J. A. C. Baum (Ed.), *The Blackwell Companion to Organizations.* Oxford: Blackwell.

DStGB. (2010). *Nr. 92 – Öffentliche Beleuchtung, Analyse, Potenziale und Beschaffung.* Deutscher Städte- und Gemeindebund.

DStGB. (2014). *BILANZ 2014 und AUSBLICK 2015 der deutschen Staedte und Gemeinden.* Berlin: Deutscher Städte- und Gemeindebund.

Duscha, M., Brischke, L.-A., Schmitt, C., Irrek, W., Ansari, E., & Meyer, C. (2013). *Marktanalyse und Marktbewertung sowie Erstellung eines Konzeptes zur Marktbeobachtung für ausgewählte Dienstleistungen im Bereich Energieeffizienz.* Retrieved from http://www.bafa.de/bfee/informationsangebote/publikationen/studien/marktanalyse_edl_ueberblick.pdf

EC. (2014). Energy Efficiency – Saving energy, saving money. Retrieved from http://ec.europa.eu/energy/en/topics/energy-efficiency

Edler, J., & Georghiou, L. (2007). Public procurement and innovation—Resurrecting the demand side. *Research Policy, 36*(7), 949–963. http://doi.org/10.1016/j.respol.2007.03.003

Edquist, C. (2011). Design of innovation policy through diagnostic analysis: identification of systemic problems (or failures). *Industrial and Corporate Change, 20*(6), 1725–1753. http://doi.org/10.1093/icc/dtr060

Edquist, C., & Zabala-Iturriagagoitia, J. M. (2012). Public Procurement for Innovation as mission-oriented innovation policy. *Research Policy, 41*(10), 1757–1769. http://doi.org/10.1016/j.respol.2012.04.022

Eichhammer, W., Ragwitz, M., & Schlomann, B. (2013). Introduction to the Special Issue: Financing Instruments to Promote Energy Efficiency and Renewables in

Times of Tight Public Budgets. *Energy & Environment*, *24*(1), 1–26. http://doi.org/10.1260/0958-305X.24.1-2.1

Eickelpasch, A., & Fritsch, M. (2005). Contests for cooperation—A new approach in German innovation policy. *Research Policy*, *34*(8), 1269–1282. http://doi.org/10.1016/j.respol.2005.02.009

Eisenhardt, K. M. (1989). Building Theories from Case Study Research. *The Academy of Management Review*, *14*(4), 532–550. http://doi.org/10.2307/258557

Energy Transition. (2013). Local, Decentralized, Innovative: Why Germany's Municipal Utilities are Right for the Energiewende. Retrieved from http://energytransition.de/2013/09/local-decentralized-innovative-why-germanys-municipal-utilities-are-right-for-the-energiewende/

Enzensberger, N., Wietschel, M., & Rentz, O. (2002). Policy instruments fostering wind energy projects—a multi-perspective evaluation approach. *Energy Policy*, *30*(9), 793–801. http://doi.org/10.1016/S0301-4215(01)00139-2

Este, P. D', Iammarino, S., Savona, M., & Tunzelmann, N. von. (2012). What hampers innovation? Revealed barriers versus deterring barriers. *Research Policy*, *41*(2), 482–488. http://doi.org/10.1016/j.respol.2011.09.008

European Commission. (2009). The 2020 climate and energy package. Retrieved from http://ec.europa.eu/clima/policies/package/index_en.htm

European Commission. (2010). Europe 2020. Retrieved from http://ec.europa.eu/europe2020/index_en.htm

Fagerberg, J., Fosaas, M., & Sapprasert, K. (2012). Innovation: Exploring the knowledge base. *Research Policy*, *41*(7), 1132–1153. http://doi.org/10.1016/j.respol.2012.03.008

Fischer, C. (2008). Emissions pricing, spillovers, and public investment in environmentally friendly technologies. *Energy Economics*, *30*(2), 487–502. http://doi.org/10.1016/j.eneco.2007.06.001

Fischer, C., & Newell, R. G. (2008). Environmental and technology policies for climate mitigation. *Journal of Environmental Economics and Management*, *55*(2), 142–162. http://doi.org/10.1016/j.jeem.2007.11.001

Flanagan, K., Uyarra, E., & Laranja, M. (2011). Reconceptualising the "policy mix" for innovation. *Research Policy*, *40*(5), 702–713. http://doi.org/10.1016/j.respol.2011.02.005

Flotow, P. von, & Schiereck, D. (2013). *Klimawandel, Finanzmärkte und Innovation* (Projektbericht). Oestrich-Winkel: Sustainable Business Institute (SBI). Retrieved from http://www.cfi21.org/cfi-themen.0.html

Flyvbjerg, B. (2006). Five Misunderstandings About Case-Study Research. *Qualitative Inquiry*, *12*(2), 219 –245. http://doi.org/10.1177/1077800405284363

Foxon, T. J., Gross, R., Chase, A., Howes, J., Arnall, A., & Anderson, D. (2005). UK innovation systems for new and renewable energy technologies: drivers, barriers and systems failures. *Energy Policy*, *33*(16), 2123–2137. http://doi.org/10.1016/j.enpol.2004.04.011

Foxon, T. J., Köhler, J., & Oughton, C. (2008). *Innovation for a Low Carbon Economy: Economic, Institutional and Management Approaches*. Edward Elgar Publishing.

Foxon, T. J., & Pearson, P. (2008). Overcoming barriers to innovation and diffusion of cleaner technologies: some features of a sustainable innovation policy regime. *Journal of Cleaner Production*, *16*(1), 148–161. http://doi.org/10.1016/j.jclepro.2007.10.011

Foxon, T. J., & Pearson, P. J. G. (2007). Towards improved policy processes for promoting innovation in renewable electricity technologies in the UK. *Energy Policy*, *35*(3), 1539–1550. http://doi.org/10.1016/j.enpol.2006.04.009

Frankfurt School-UNEP Centre, & BNEF. (2014). *Global Trends in Renewable Energy Investment 2014*. Frankfurt am Main. Retrieved from http://www.fs-unep-centre.org

Freeman, C. (1996). The greening of technology and models of innovation. *Technological Forecasting and Social Change*, *53*(1), 27–39. http://doi.org/10.1016/0040-1625(96)00060-1

Friebe, C. A., Flotow, P. von, & Täube, F. A. (2013). Exploring the link between products and services in low-income markets—Evidence from solar home systems. *Energy Policy*, *52*, 760–769. http://doi.org/10.1016/j.enpol.2012.10.038

Friebe, C. A., Flotow, P. von, & Täube, F. A. (2014). Exploring technology diffusion in emerging markets – the role of public policy for wind energy. *Energy Policy*, *70*, 217–226. http://doi.org/10.1016/j.enpol.2014.03.016

Frondel, M., Horbach, J., & Rennings, K. (2007). End-of-pipe or cleaner production? An empirical comparison of environmental innovation decisions across OECD countries. *Business Strategy and the Environment*, *16*(8), 571–584. http://doi.org/10.1002/bse.496

Gadamer, H. (1993). *Truth and Method*. New York: Continuum.

Gallagher, K. S., Holdren, J. P., & Sagar, A. D. (2006). Energy-Technology Innovation. *Annual Review of Environment and Resources*, *31*, 193–237. http://doi.org/10.1146/annurev.energy.30.050504.144321

Geels, F. W. (2010). Ontologies, socio-technical transitions (to sustainability), and the multi-level perspective. *Research Policy*, *39*(4), 495–510. http://doi.org/10.1016/j.respol.2010.01.022

Gee, S., & McMeekin, A. (2011). Eco-Innovation Systems and Problem Sequences: The Contrasting Cases of US and Brazilian Biofuels. *Industry and Innovation*, *18*(3), 301–315. http://doi.org/10.1080/13662716.2011.561029

Godoe, H., & Nygaard, S. (2006). System failure, innovation policy and patents: Fuel cells and related hydrogen technology in Norway 1990–2002. *Energy Policy, 34*(13), 1697–1708. http://doi.org/10.1016/j.enpol.2004.12.016

Goulder, L. H., & Mathai, K. (2000). Optimal CO2 Abatement in the Presence of Induced Technological Change. *Journal of Environmental Economics and Management, 39*(1), 1–38. http://doi.org/10.1006/jeem.1999.1089

Griliches, Z. (1992). The search for R&D spillovers. *National Bureau of Economic Research Working Paper Series*, No. 3768. Retrieved from http://www.nber.org/papers/w3768

Guerzoni, M., & Raiteri, E. (2014). Demand-side vs. supply-side technology policies: Hidden treatment and new empirical evidence on the policy mix. *Research Policy.* http://doi.org/10.1016/j.respol.2014.10.009

Haas, R., Nakicenovic, N., Ajanovic, A., Faber, T., Kranzl, L., Müller, A., & Resch, G. (2008). Towards sustainability of energy systems: A primer on how to apply the concept of energy services to identify necessary trends and policies. *Energy Policy, 36*(11), 4012–4021. http://doi.org/10.1016/j.enpol.2008.06.028

Hair, J. (2010). *Multivariate data analysis : a global perspective.* (7th ed. /). Upper Saddle River N.J., London: Pearson Education.

Haley, U. C. V., & Schuler, D. A. (2011). Government Policy and Firm Strategy in the Solar Photovoltaic Industry. *California Management Review, 54*(1), 17–38. http://doi.org/10.1525/cmr.2011.54.1.17

Hall, B. H. (2002). The Financing of Research and Development. *National Bureau of Economic Research Working Paper Series, No. 8773.* Retrieved from http://www.nber.org/papers/w8773

Hall, B. H., & Helmers, C. (2011). *Innovation and Diffusion of Clean/Green Technology: Can Patent Commons Help?* (Working Paper No. 16920). National Bureau of Economic Research. Retrieved from http://www.nber.org/papers/w16920

Hall, B. H., & Lerner, J. (2010). The Financing of R&D and Innovation. In B. H. Hall & N. Rosenberg (Eds.), *Handbook of The Economics of Innovation, Vol. 1* (Vol. Volume 1, pp. 609–639). North-Holland.

Hall, J., & Kerr, R. (2003). Innovation dynamics and environmental technologies: the emergence of fuel cell technology. *Journal of Cleaner Production, 11*(4), 459–471. http://doi.org/10.1016/S0959-6526(02)00067-7

Hannon, M. J., & Bolton, R. (2014). UK Local Authority engagement with the Energy Service Company (ESCo) model: Key characteristics, benefits, limitations and considerations. *Energy Policy.* http://doi.org/10.1016/j.enpol.2014.11.016

Hannon, M. J., Foxon, T. J., & Gale, W. F. (2013). The co-evolutionary relationship between Energy Service Companies and the UK energy system: Implications for a low-carbon transition. *Energy Policy*, *61*, 1031–1045. http://doi.org/10.1016/j.enpol.2013.06.009

Hargadon, A. (2010). Technology policy and global warming: Why new innovation models are needed. *Research Policy*, *39*(8), 1024–1026. http://doi.org/10.1016/j.respol.2010.05.009

Harmelink, M., Voogt, M., & Cremer, C. (2006). Analysing the effectiveness of renewable energy supporting policies in the European Union. *Energy Policy*, *34*(3), 343–351. http://doi.org/10.1016/j.enpol.2004.08.031

Hart, C. (1998). *Doing a literature review: Releasing the social science research imagination*. Sage.

Hartmann, A., Roehrich, J., Frederiksen, L., & Davies, A. (2014). Procuring complex performance: the transition process in public infrastructure. *International Journal of Operations & Production Management*, *34*(2), 174–194. http://doi.org/10.1108/IJOPM-01-2011-0032

Hekkert, M. P., & Negro, S. O. (2009). Functions of innovation systems as a framework to understand sustainable technological change: Empirical evidence for earlier claims. *Technological Forecasting and Social Change*, *76*(4), 584–594. http://doi.org/10.1016/j.techfore.2008.04.013

Hekkert, M. P., Suurs, R. A. A., Negro, S. O., Kuhlmann, S., & Smits, R. E. H. M. (2007). Functions of innovation systems: A new approach for analysing technological change. *Technological Forecasting and Social Change*, *74*(4), 413–432. http://doi.org/10.1016/j.techfore.2006.03.002

Helle, C. (1997). On energy efficiency-related product strategies—illustrated and analysed using contracting approaches in Germany as an example. *Utilities Policy*, *6*(1), 75–85. http://doi.org/10.1016/S0957-1787(96)00011-2

Helm, D. (2002). Energy policy: security of supply, sustainability and competition. *Energy Policy*, *30*(3), 173–184. http://doi.org/10.1016/S0301-4215(01)00141-0

Hendricks, B. (2014). Klimaziele müssen weiter verfolgt werden. Retrieved from http://www.bundesregierung.de/Content/DE/Interview/2014/01/2014-01-16-hendricks-dlf.html

Hendry, C., Harborne, P., & Brown, J. (2010). So what do innovating companies really get from publicly funded demonstration projects and trials? innovation lessons from solar photovoltaics and wind. *Energy Policy*, *38*(8), 4507–4519. http://doi.org/10.1016/j.enpol.2010.04.005

Henriot, A. (2013). Financing investment in the European electricity transmission network: Consequences on long-term sustainability of the TSOs financial structure. *Energy Policy*, *62*, 821–829. http://doi.org/10.1016/j.enpol.2013.07.011

Herring, H. (2006). Energy efficiency—a critical view. *Energy*, *31*(1), 10–20. http://doi.org/10.1016/j.energy.2004.04.055

Hockerts, K., & Wüstenhagen, R. (2010). Greening Goliaths versus emerging Davids – Theorizing about the role of incumbents and new entrants in sustainable entrepreneurship. *Journal of Business Venturing*, *25*(5), 481–492. http://doi.org/10.1016/j.jbusvent.2009.07.005

Holmstrom, B., & Tirole, J. (1997). Financial Intermediation, Loanable Funds, and The Real Sector. *The Quarterly Journal of Economics*, *112*(3), 663–691. http://doi.org/10.1162/003355397555316

Hoppe, H. C., & Ozdenoren, E. (2005). Intermediation in innovation. *International Journal of Industrial Organization*, *23*(5–6), 483–503. http://doi.org/10.1016/j.ijindorg.2005.03.003

Horbach, J., Chen, Q., Rennings, K., & Vögele, S. (2013). Do lead markets for clean coal technology follow market demand? A case study for China, Germany, Japan and the US. *Environmental Innovation and Societal Transitions*. http://doi.org/10.1016/j.eist.2013.08.002

Horbach, J., Rammer, C., & Rennings, K. (2012). Determinants of eco-innovations by type of environmental impact — The role of regulatory push/pull, technology push and market pull. *Ecological Economics*, *78*, 112–122. http://doi.org/10.1016/j.ecolecon.2012.04.005

Howells, J. (2006). Intermediation and the role of intermediaries in innovation. *Research Policy*, *35*(5), 715–728. http://doi.org/10.1016/j.respol.2006.03.005

Huberty, M., & Zysman, J. (2010). An energy system transformation: Framing research choices for the climate challenge. *Research Policy*, *39*(8), 1027–1029. http://doi.org/10.1016/j.respol.2010.05.010

Hunter, J. E., & Schmidt, F. L. (2004). *Methods of meta-analysis: correcting error and bias in research findings*. SAGE.

Hypko, P., Tilebein, M., & Gleich, R. (2010a). Benefits and uncertainties of performance-based contracting in manufacturing industries: An agency theory perspective. *Journal of Service Management*, *21*(4), 460–489. http://doi.org/10.1108/09564231011066114

Hypko, P., Tilebein, M., & Gleich, R. (2010b). Clarifying the concept of performance-based contracting in manufacturing industries: A research synthesis. *Journal of Service Management*, *21*(5), 625–655. http://doi.org/10.1108/09564231011079075

IEA. (2013a). *Solid State Lighting Annex: 2013 Interlaboratory Comparison*. International Energy Agency. Retrieved from http://ssl.iea-4e.org/task-2-ssl-testing/2013-ic-final-report

IEA. (2013b). *Tracking Clean Energy Progress 2013*. Paris: IEA. Retrieved from http://www.iea.org/etp/tracking/

IEA. (2014a). *Capturing the Multiple Benefits of Energy Efficiency*. Paris: International Energy Agency.

IEA. (2014b). *Energy Efficiency Market Report 2014*. International Energy Agency. Retrieved from http://www.iea.org/Textbase/npsum/EEMR2014SUM.pdf

IEA. (2014c). IEA FAQ Renewable Energy. Retrieved from http://www.iea.org/aboutus/faqs/renewableenergy/

IPCC. (2014). *Climate Change 2014: Mitigation of climate change – IPCC Working Group III Contribution to AR5* (p. 33). Berlin: Int.

Iyer, G., Hultman, N., Eom, J., McJeon, H., Patel, P., & Clarke, L. (2013). Diffusion of low-carbon technologies and the feasibility of long-term climate targets. *Technological Forecasting and Social Change*. http://doi.org/10.1016/j.techfore.2013.08.025

Jackson, J. (2010). Promoting energy efficiency investments with risk management decision tools. *Energy Policy, 38*(8), 3865–3873. http://doi.org/10.1016/j.enpol.2010.03.006

Jacobsson, S., & Bergek, A. (2004). Transforming the energy sector: the evolution of technological systems in renewable energy technology. *Industrial and Corporate Change, 13*(5), 815–849. http://doi.org/10.1093/icc/dth032

Jacobsson, S., & Bergek, A. (2011). Innovation system analyses and sustainability transitions: Contributions and suggestions for research. *Environmental Innovation and Societal Transitions, 1*(1), 41–57. http://doi.org/10.1016/j.eist.2011.04.006

Jacobsson, S., Bergek, A., Finon, D., Lauber, V., Mitchell, C., Toke, D., & Verbruggen, A. (2009). EU renewable energy support policy: Faith or facts? *Energy Policy, 37*(6), 2143–2146. http://doi.org/10.1016/j.enpol.2009.02.043

Jacobsson, S., & Karltorp, K. (2013). Mechanisms blocking the dynamics of the European offshore wind energy innovation system – Challenges for policy intervention. *Energy Policy, 63*, 1182–1195. http://doi.org/10.1016/j.enpol.2013.08.077

Jacobsson, S., & Lauber, V. (2006). The politics and policy of energy system transformation—explaining the German diffusion of renewable energy technology. *Energy Policy, 34*(3), 256–276. http://doi.org/10.1016/j.enpol.2004.08.029

Jaffe, A. B., Newell, R. G., & Stavins, R. N. (2002). Environmental policy and technological change. *Environmental and Resource Economics, 22*(1), 41–70. http://doi.org/10.1023/A:1015519401088

Jaffe, A. B., Newell, R. G., & Stavins, R. N. (2005). A tale of two market failures: Technology and environmental policy. *Ecological Economics, 54*(2–3), 164–174. http://doi.org/10.1016/j.ecolecon.2004.12.027

Jaffe, A. B., & Stavins, R. N. (1994). The energy paradox and the diffusion of conservation technology. *Resource and Energy Economics*, *16*(2), 91–122. http://doi.org/10.1016/0928-7655(94)90001-9

Jaffe, A. B., & Stavins, R. N. (1995). Dynamic Incentives of Environmental Regulations: The Effects of Alternative Policy Instruments on Technology Diffusion. *Journal of Environmental Economics and Management*, *29*(3), S43–S63. http://doi.org/10.1006/jeem.1995.1060

Jakeman, G., Hanslow, K., Hinchy, M., Fisher, B. S., & Woffenden, K. (2004). Induced innovations and climate change policy. *Energy Economics*, *26*(6), 937–960. http://doi.org/10.1016/j.eneco.2004.09.002

Jefferson, M. (2008). Accelerating the transition to sustainable energy systems. *Energy Policy*, *36*(11), 4116–4125. http://doi.org/10.1016/j.enpol.2008.06.020

Jenner, S., Groba, F., & Indvik, J. (2013). Assessing the strength and effectiveness of renewable electricity feed-in tariffs in European Union countries. *Energy Policy*, *52*, 385–401. http://doi.org/10.1016/j.enpol.2012.09.046

Jensen, S. G., & Skytte, K. (2002). Interactions between the power and green certificate markets. *Energy Policy*, *30*(5), 425–435. http://doi.org/10.1016/S0301-4215(01)00111-2

Johnstone, N., Haščič, I., & Popp, D. (2010). Renewable Energy Policies and Technological Innovation: Evidence Based on Patent Counts. *Environmental and Resource Economics*, *45*(1), 133–155. http://doi.org/10.1007/s10640-009-9309-1

Katzy, B., Turgut, E., Holzmann, T., & Sailer, K. (2013). Innovation intermediaries: a process view on open innovation coordination. *Technology Analysis & Strategic Management*, *25*(3), 295–309. http://doi.org/10.1080/09537325.2013.764982

Kemp, R., & Oltra, V. (2011). Research Insights and Challenges on Eco-Innovation Dynamics. *Industry and Innovation*, *18*(3), 249–253. http://doi.org/10.1080/13662716.2011.562399

Kenney, M., & Hargadon, A. (2012). Misguided Policy? *California Management Review*, *54*(2), 118–139. http://doi.org/10.1525/cmr.2012.54.2.118

Kern, F., & Smith, A. (2008). Restructuring energy systems for sustainability? Energy transition policy in the Netherlands. *Energy Policy*, *36*(11), 4093–4103. http://doi.org/10.1016/j.enpol.2008.06.018

Kimura, O. (2010). Public R&D and commercialization of energy-efficient technology: A case study of Japanese projects. *Energy Policy*, *38*(11), 7358–7369. http://doi.org/10.1016/j.enpol.2010.08.012

Kivimaa, P. (2014). Government-affiliated intermediary organisations as actors in system-level transitions. *Research Policy*. http://doi.org/10.1016/j.respol.2014.02.007

Kivimaa, P., & Mickwitz, P. (2006). The challenge of greening technologies—Environmental policy integration in Finnish technology policies. *Research Policy*, 35(5), 729–744. http://doi.org/10.1016/j.respol.2006.03.006

Kleer, R. (2010). Government R&D subsidies as a signal for private investors. *Research Policy*, 39(10), 1361–1374. http://doi.org/10.1016/j.respol.2010.08.001

Klein Woolthuis, R., Lankhuizen, M., & Gilsing, V. (2005). A system failure framework for innovation policy design. *Technovation*, 25(6), 609–619. http://doi.org/10.1016/j.technovation.2003.11.002

Klerkx, L., Álvarez, R., & Campusano, R. (2014). The emergence and functioning of innovation intermediaries in maturing innovation systems: the case of Chile. *Innovation and Development*, 1–19. http://doi.org/10.1080/2157930X.2014.921268

Klerkx, L., & Leeuwis, C. (2009). Establishment and embedding of innovation brokers at different innovation system levels: Insights from the Dutch agricultural sector. *Technological Forecasting and Social Change*, 76(6), 849–860. http://doi.org/10.1016/j.techfore.2008.10.001

Kley, F., Lerch, C., & Dallinger, D. (2011). New business models for electric cars—A holistic approach. *Energy Policy*, 39(6), 3392–3403. http://doi.org/10.1016/j.enpol.2011.03.036

Kobos, P. H., Erickson, J. D., & Drennen, T. E. (2006). Technological learning and renewable energy costs: implications for US renewable energy policy. *Energy Policy*, 34(13), 1645–1658. http://doi.org/10.1016/j.enpol.2004.12.008

Köhler, J., Wietschel, M., Whitmarsh, L., Keles, D., & Schade, W. (2010). Infrastructure investment for a transition to hydrogen automobiles. *Technological Forecasting and Social Change*, 77(8), 1237–1248. http://doi.org/10.1016/j.techfore.2010.03.010

Komor, P., & Bazilian, M. (2005). Renewable energy policy goals, programs, and technologies. *Energy Policy*, 33(14), 1873–1881. http://doi.org/10.1016/j.enpol.2004.03.003

Kverndokk, S., Rosendahl, K. E., & Rutherford, T. F. (2004). Climate Policies and Induced Technological Change: Which to Choose, the Carrot or the Stick? *Environmental and Resource Economics*, 27(1), 21–41. http://doi.org/10.1023/B:EARE.0000016787.53575.39

Lazonick, W., & Mazzucato, M. (2013). The risk-reward nexus in the innovation-inequality relationship: who takes the risks? Who gets the rewards? *Industrial and Corporate Change*, 22(4), 1093–1128. http://doi.org/10.1093/icc/dtt019

Leete, S., Xu, J., & Wheeler, D. (2013). Investment barriers and incentives for marine renewable energy in the UK: An analysis of investor preferences. *Energy Policy*, 60, 866–875. http://doi.org/10.1016/j.enpol.2013.05.011

Leitner, A., Wehrmeyer, W., & France, C. (2010). The impact of regulation and policy on radical eco-innovation: The need for a new understanding. *Management Research Review*, *33*(11), 1022–1041. http://doi.org/10.1108/01409 171011085877

Lerner, J. (1999). The Government as Venture Capitalist: The Long-Run Impact of the SBIR Program. *The Journal of Business*, *72*(3), 285–318. http://www.jstor.org/stable/10.1086/209616

Lesser, J., & Su, X. (2008). Design of an economically efficient feed-in tariff structure for renewable energy development. *Energy Policy*, *36*, 981–990. http://doi.org/10.1016/j.enpol.2007.11.007

Lewis, J. I., & Wiser, R. H. (2007). Fostering a renewable energy technology industry: An international comparison of wind industry policy support mechanisms. *Energy Policy*, *35*(3), 1844–1857.

Link, A. N., & Scott, J. T. (2010). Government as entrepreneur: Evaluating the commercialization success of SBIR projects. *Research Policy*, *39*(5), 589–601. http://doi.org/16/j.respol.2010.02.006

Loiter, J. M., & Norberg-Bohm, V. (1999). Technology policy and renewable energy: public roles in the development of new energy technologies. *Energy Policy*, *27*(2), 85–97. http://doi.org/10.1016/S0301-4215(99)00013-0

Luiten, E., & Blok, K. (2003). Stimulating R&D of industrial energy-efficient technology; the effect of government intervention on the development of strip casting technology. *Energy Policy*, *31*(13), 1339–1356. http://doi.org/10.1016/S0301-4215(02)00194-5

Luiten, E., van Lente, H., & Blok, K. (2006). Slow technologies and government intervention: Energy efficiency in industrial process technologies. *Technovation*, *26*(9), 1029–1044. http://doi.org/10.1016/j.technovation.2005.10.004

Lüthi, S., & Prässler, T. (2011). Analyzing policy support instruments and regulatory risk factors for wind energy deployment—A developers' perspective. *Energy Policy*, *39*(9), 4876–4892. http://doi.org/10.1016/j.enpol.2011.06.029

Lüthi, S., & Wüstenhagen, R. (2012). The price of policy risk — Empirical insights from choice experiments with European photovoltaic project developers. *Energy Economics*, *34*(4), 1001–1011. http://doi.org/10.1016/j.eneco.2011.08.007

Malerba, F. (2002). Sectoral systems of innovation and production. *Research Policy*, *31*(2), 247–264. http://doi.org/10.1016/S0048-7333(01)00139-1

Mantere, S. (2008). Role Expectations and Middle Manager Strategic Agency. *Journal of Management Studies*, *45*(2), 294–316. http://doi.org/10.1111/j.1467-6486.2007.00744.x

Marcucci, A., & Turton, H. (2013). Induced technological change in moderate and fragmented climate change mitigation regimes. *Technological Forecasting and Social Change*. http://doi.org/10.1016/j.techfore.2013.10.027

Marcus, A., Malen, J., & Ellis, S. (2013). The Promise and Pitfalls of Venture Capital as an Asset Class for Clean Energy Investment Research Questions for Organization and Natural Environment Scholars. *Organization & Environment, 26*(1), 31–60. http://doi.org/10.1177/1086026612474956

Marino, A., Bertoldi, P., & Rezessy, S. (2010). *Energy Service Companies Market in Europe – Status Report 2010 -*. Retrieved from http://publications.jrc.ec.europa.eu/repository/handle/111111111/15108

Marino, A., Bertoldi, P., Rezessy, S., & Boza-Kiss, B. (2011). A snapshot of the European energy service market in 2010 and policy recommendations to foster a further market development. *Energy Policy, 39*(10), 6190–6198. http://doi.org/10.1016/j.enpol.2011.07.019

Markard, J., Raven, R., & Truffer, B. (2012). Sustainability transitions: An emerging field of research and its prospects. *Research Policy, 41*(6), 955–967. http://doi.org/10.1016/j.respol.2012.02.013

Markard, J., & Truffer, B. (2008). Technological innovation systems and the multi-level perspective: Towards an integrated framework. *Research Policy, 37*(4), 596–615. http://doi.org/10.1016/j.respol.2008.01.004

Marques, A. C., & Fuinhas, J. A. (2011). Do energy efficiency measures promote the use of renewable sources? *Environmental Science & Policy, 14*(4), 471–481. http://doi.org/10.1016/j.envsci.2011.02.001

Marques, A. C., & Fuinhas, J. A. (2012a). Are public policies towards renewables successful? Evidence from European countries. *Renewable Energy, 44*, 109–118. http://doi.org/10.1016/j.renene.2012.01.007

Marques, A. C., & Fuinhas, J. A. (2012b). Is renewable energy effective in promoting growth? *Energy Policy, 46*, 434–442. http://doi.org/10.1016/j.enpol.2012.04.006

Marques, A. C., Fuinhas, J. A., & Pires Manso, J. R. (2010). Motivations driving renewable energy in European countries: A panel data approach. *Energy Policy, 38*(11), 6877–6885. http://doi.org/10.1016/j.enpol.2010.07.003

Martin, B. R. (2012a). Innovation Studies: Challenging the Boundaries. In *Lundvall Symposium on the Future of Innovation Studies, 16-17 February 2012, Aalborg University.*

Martin, B. R. (2012b). The evolution of science policy and innovation studies. *Research Policy, 41*(7), 1219–1239. http://doi.org/10.1016/j.respol.2012.03.012

Martin, B. R., Nightingale, P., & Yegros-Yegros, A. (2012). Science and technology studies: Exploring the knowledge base. *Research Policy, 41*(7), 1182–1204. http://doi.org/10.1016/j.respol.2012.03.010

Masini, A., & Menichetti, E. (2012). The impact of behavioural factors in the renewable energy investment decision making process: Conceptual framework and empirical findings. *Energy Policy, 40*, 28–38. http://doi.org/10.1016/j.enpol. 2010.06.062

Mathews, J. A., Kidney, S., Mallon, K., & Hughes, M. (2010). Mobilizing private finance to drive an energy industrial revolution. *Energy Policy, 38*(7), 3263–3265. http://doi.org/10.1016/j.enpol.2010.02.030

Maxwell, D., Sheate, W., & van der Vorst, R. (2006). Functional and systems aspects of the sustainable product and service development approach for industry. *Journal of Cleaner Production, 14*, 1466–1479. http://doi.org/10.1016/j. jclepro.2006.01.028

Maxwell, D., & van der Vorst, R. (2003). Developing sustainable products and services. *Journal of Cleaner Production, 11*, 883–895. http://doi.org/10.1016/ S0959-6526(02)00164-6

Mazzucato, M. (2013). *The Entrepreneurial State – Debunking Private Vs. Public Sector Myths*. Anthem Press.

Mazzucato, M., & Perez, C. (2014). Innovation as Growth Policy: The challenge for Europe. *SPRU Working Paper Series, SWPS 2014-13*. Retrieved from http://www.sussex.ac.uk/spru/research/swps

McKinsey. (2012). *Lighting the way: Perspectives on the global lighting market (Second Edition)*. McKinsey&Company. Retrieved from http://www.mckinsey. com/~/media/McKinsey/dotcom/client_service/Automotive%20and%20Assembly/Lighting_the_way_Perspectives_on_global_lighting_market_2012. ashx

Meadows, D. H., Goldsmith, E. I., & Meadows, P. (1972). *The limits to growth* (Vol. 381). Earth Island Limited London.

Meadows, D. H., Randers, J., & Meadows, D. L. (2004). *Limits to Growth: The 30-Year Update*. Chelsea Green Publishing.

Menanteau, P., Finon, D., & Lamy, M.-L. (2003). Prices versus quantities: choosing policies for promoting the development of renewable energy. *Energy Policy, 31*(8), 799–812. http://doi.org/10.1016/S0301-4215(02)00133-7

Menz, F. C., & Vachon, S. (2006). The effectiveness of different policy regimes for promoting wind power: Experiences from the states. *Energy Policy, 34*(14), 1786–1796. http://doi.org/10.1016/j.enpol.2004.12.018

Meuleman, M., & De Maeseneire, W. (2012). Do R&D subsidies affect SMEs' access to external financing? *Research Policy, 41*(3), 580–591. http://doi. org/10.1016/j.respol.2012.01.001

Mickwitz, P., Hyvättinen, H., & Kivimaa, P. (2008). The role of policy instruments in the innovation and diffusion of environmentally friendlier technologies: popular claims versus case study experiences. *Journal of Cleaner Production, 16*(1, Supplement 1), S162–S170. http://doi.org/10.1016/j.jclepro.2007.10.012

Migendt, M., Schock, F., Flotow, V. von, Täube, F. A., & Polzin, F. (2014). *Private Equity in Clean Technology – An Exploratory Study of the Investment-Policy Nexus* (SSRN Scholarly Paper). Rochester, NY: Social Science Research Network.

Miles, M. B., & Huberman, A. M. (1999). *Qualitative data analysis.* Sage Publ.

Mills, B., & Schleich, J. (2014). Household transitions to energy efficient lighting. *Energy Economics, 46,* 151–160. http://doi.org/10.1016/j.eneco.2014.08.022

Mitchell, C., Bauknecht, D., & Connor, P. M. (2006). Effectiveness through risk reduction: a comparison of the renewable obligation in England and Wales and the feed-in system in Germany. *Energy Policy, 34*(3), 297–305. http://doi.org/10.1016/j.enpol.2004.08.004

Montalvo, C. (2008). General wisdom concerning the factors affecting the adoption of cleaner technologies: a survey 1990–2007. *Journal of Cleaner Production, 16*(1, Supplement 1), S7–S13. http://doi.org/10.1016/j.jclepro.2007.10.002

Moran-Ellis, J., Alexander, V. D., Cronin, A., Dickinson, M., Fielding, J., Sleney, J., & Thomas, H. (2006). Triangulation and integration: processes, claims and implications. *Qualitative Research, 6*(1), 45–59. http://doi.org/10.1177/1468794106058870

Mowery, D. C., Nelson, R. R., & Martin, B. R. (2010). Technology policy and global warming: Why new policy models are needed (or why putting new wine in old bottles won't work). *Research Policy, 39,* 1011–1023. http://doi.org/10.1016/j.respol.2010.05.008

Müller, S., Brown, A., & Ölz, S. (2011). Renewable energy: policy considerations for deploying renewables. *International Energy Agency.*

Myers, S. C., & Majluf, N. S. (1984). Corporate financing and investment decisions when firms have information that investors do not have. *Journal of Financial Economics, 13*(2), 187–221. http://doi.org/10.1016/0304-405X(84)90023-0

Neij, L., & Åstrand, K. (2006). Outcome indicators for the evaluation of energy policy instruments and technical change. *Energy Policy, 34*(17), 2662–2676. http://doi.org/10.1016/j.enpol.2005.03.012

Nemet, G. F. (2012). Subsidies for New Technologies and Knowledge Spillovers from Learning by Doing. *Journal of Policy Analysis and Management, 31*(3), 601–622. http://doi.org/10.1002/pam.21643

Nemet, G. F., & Kammen, D. M. (2007). U.S. energy research and development: Declining investment, increasing need, and the feasibility of expansion. *Energy Policy, 35*(1), 746–755. http://doi.org/10.1016/j.enpol.2005.12.012

Nesta, L., Vona, F., & Nicolli, F. (2014). Environmental policies, competition and innovation in renewable energy. *Journal of Environmental Economics and Management, 67*(3), 396–411. http://doi.org/10.1016/j.jeem.2014.01.001

Newell, R. G., Jaffe, A. B., & Stavins, R. N. (2006). The effects of economic and policy incentives on carbon mitigation technologies. *Energy Economics, 28*(5–6), 563–578. http://doi.org/10.1016/j.eneco.2006.07.004

Nightingale, P., Murray, G., Cowling, M., Baden-Fuller, C., Mason, C., Siepel, J., … Dannreuther, C. (2009). From funding gaps to thin markets: UK Government support for early-stage venture capital. Retrieved from https://eric.exeter.ac.uk/repository/bitstream/handle/10036/106659/Thin-Markets-v9.pdf?sequence=2

Nill, J., & Kemp, R. (2009). Evolutionary approaches for sustainable innovation policies: From niche to paradigm? *Research Policy, 38*(4), 668–680. http://doi.org/10.1016/j.respol.2009.01.011

Noailly, J., & Batrakova, S. (2010). Stimulating energy-efficient innovations in the Dutch building sector: Empirical evidence from patent counts and policy lessons. *Energy Policy, 38*(12), 7803–7817. http://doi.org/10.1016/j.enpol.2010.08.040

NREL. (2013). *2012 Renewable Energy Data Book* (p. 128). National Renewable Energy Laboratory. Retrieved from http://www.nrel.gov/docs/fy14osti/60197.pdf

Oakey, R. P. (2003). Funding innovation and growth in UK new technology-based firms: Some observations on contributions from the public and private sectors. *Venture Capital: An International Journal of Entrepreneurial Finance, 5*(2), 161–179. http://doi.org/10.1080/1369106032000097049

OECD. (2009). *Overcoming the Crisis and Beyond*. OECD. Retrieved from http://www1.oecd.org/env/43176103.pdf

OECD. (2013). *OECD factbook: economic, environmental and social statistics*. OECD.

Olmos, L., Ruester, S., & Liong, S.-J. (2012). On the selection of financing instruments to push the development of new technologies: Application to clean energy technologies. *Energy Policy, 43*, 252–266. http://doi.org/10.1016/j.enpol.2012.01.001

Oltra, V., & Jean, M. S. (2005). The dynamics of environmental innovations: three stylised trajectories of clean technology. *Economics of Innovation and New Technology, 14*, 189–212. http://doi.org/10.1080/1043859042000226202

O'Sullivan, M. (2006). Finance and Innovation. In J. Fagerberg, D. C. Mowery, & R. R. Nelson (Eds.), *The Oxford handbook of innovation* (pp. 240–265). Oxford: Oxford University Press.

Painuly, J. P., Park, H., Lee, M.-K., & Noh, J. (2003). Promoting energy efficiency financing and ESCOs in developing countries: mechanisms and barriers.

Journal of Cleaner Production, 11, 659–665. http://doi.org/10.1016/S0959-6526(02)00111-7

Pätäri, S., & Sinkkonen, K. (2014). Energy Service Companies and Energy Performance Contracting: is there a need to renew the business model? Insights from a Delphi study. *Journal of Cleaner Production, 66,* 264–271. http://doi.org/10.1016/j.jclepro.2013.10.017

Patton, M. Q. (2002). *Qualitative research and evaluation methods.* SAGE.

Pellegrino, G., & Savona, M. (2013). Is money all? Financing versus knowledge and demand constraints to innovation. *SPRU Working Paper Series, SWPS 2013-01.*

Perez, C. (2002). *Technological revolutions and financial capital: The dynamics of bubbles and golden ages.* Edward Elgar Publishing.

Perez, C. (2004). Technological revolutions, paradigm shifts and socio-institutional change. *Globalization, Economic Development and Inequality: An Alternative Perspective, Cheltenham, UK: Edward Elgar,* 217–242.

Perez, C. (2009). The double bubble at the turn of the century: technological roots and structural implications. *Cambridge Journal of Economics, 33*(4), 779–805. http://doi.org/10.1093/cje/bep028

Perez, C. (2013). Unleashing a golden age after the financial collapse: Drawing lessons from history. *Environmental Innovation and Societal Transitions, 6,* 9–23. http://doi.org/10.1016/j.eist.2012.12.004

Pint, E. M., & Baldwin, L. H. (1997). *Strategic Sourcing Theory and Evidence from Economics and Business Management.*

Pittaway, L., Robertson, M., Munir, K., Denyer, D., & Neely, A. (2004). Networking and innovation: a systematic review of the evidence. *International Journal of Management Reviews, 5*(3/4), 137–168. http://doi.org/10.1111/j.1460-8545.2004.00101.x

Polkinghorne, D. (1988). *Narrative knowing and the human sciences.* State Univ of New York Press.

Popp, D. (2006). Innovation in climate policy models: Implementing lessons from the economics of R&D. *Energy Economics, 28*(5–6), 596–609. http://doi.org/10.1016/j.eneco.2006.05.007

Popp, D. (2010). *Innovation and climate policy.* National Bureau of Economic Research Cambridge, Mass., USA.

Popp, D. (2012). The Role of Technological Change in Green Growth. *Policy Research Working Paper, 6239.*

Popp, D., Hafner, T., & Johnstone, N. (2011). Environmental policy vs. public pressure: Innovation and diffusion of alternative bleaching technologies in the pulp industry. *Research Policy, 40*(9), 1253–1268. http://doi.org/10.1016/j.respol.2011.05.018

Popp, D., Hascic, I., & Medhi, N. (2011). Technology and the diffusion of renewable energy. *Energy Economics, 33*(4), 648–662. http://doi.org/10.1016/j.eneco.2010.08.007

Quirion, P. (2010). Complying with the Kyoto Protocol under uncertainty: Taxes or tradable permits? *Energy Policy, 38*(9), 5166–5173. http://doi.org/10.1016/j.enpol.2010.04.054

Radulovic, D., Skok, S., & Kirincic, V. (2011). Energy efficiency public lighting management in the cities. *Energy, 36*(4), 1908–1915. http://doi.org/10.1016/j.energy.2010.10.016

Ragwitz, M., Held, A., Pfluger, B., Nuemann, K., Resch, G., Faber, T., & Wiser, R. (2008). *Assessment of the effectiveness of renewable energy support schemes in the BRICS countries.* Fraunhofer Institute Systems and Innovation Research, Karlsruhe/ Energy Economics Group, Vienna University of Technology, Vienna/ Lawrence Berkeley National Laboratory, Berkeley.

Randjelovic, J., O'Rourke, A. R., & Orsato, R. J. (2003). The emergence of green venture capital. *Business Strategy and the Environment, 12*(4), 240–253. http://doi.org/10.1002/bse.361

Reed, W. R., & Ye, H. (2011). Which panel data estimator should I use? *Applied Economics, 43*(8), 985–1000. http://doi.org/10.1080/00036840802600087

Rennings, K. (2000). Redefining innovation: eco-innovation research and the contribution from ecological economics. *Ecological Economics, 32*(2), 319–332. http://doi.org/10.1016/S0921-8009(99)00112-3

Rennings, K., & Rammer, C. (2011). The Impact of Regulation-Driven Environmental Innovation on Innovation Success and Firm Performance. *Industry and Innovation, 18*(3), 255–283. http://doi.org/10.1080/13662716.2011.561027

Repullo, R., & Suarez, J. (2000). Entrepreneurial moral hazard and bank monitoring: A model of the credit channel. *European Economic Review, 44*(10), 1931–1950. http://doi.org/10.1016/S0014-2921(99)00069-0

Roehrich, J., & Lewis, M. (2014). Procuring complex performance: implications for exchange governance complexity. *International Journal of Operations & Production Management, 34*(2), 221–241. http://doi.org/10.1108/IJOPM-01-2011-0024

Roehrich, J., Lewis, M., & George, G. (2014). Are public–private partnerships a healthy option? A systematic literature review. *Social Science & Medicine, 113*, 110–119. http://doi.org/10.1016/j.socscimed.2014.03.037

Rogers, E. (1995). *Diffusion of Innovations.*

Rogge, K. S., & Hoffmann, V. H. (2010). The impact of the EU ETS on the sectoral innovation system for power generation technologies – Findings for Germany. *Energy Policy*, *38*(12), 7639–7652. http://doi.org/10.1016/j.enpol.2010.07.047

Rogge, K. S., Schneider, M., & Hoffmann, V. H. (2011). The innovation impact of the EU Emission Trading System — Findings of company case studies in the German power sector. *Ecological Economics*, *70*, 513–523. http://doi.org/10.1016/j.ecolecon.2010.09.032

Rosen, R. A., & Guenther, E. (2014). The economics of mitigating climate change: What can we know? *Technological Forecasting and Social Change*. http://doi.org/10.1016/j.techfore.2014.01.013

Saint Jean, M. (2008). Polluting emissions standards and clean technology trajectories under competitive selection and supply chain pressure. *Journal of Cleaner Production*, *16*(1, Supplement 1), S113–S123. http://doi.org/10.1016/j.jclepro.2007.10.009

Samila, S., & Sorenson, O. (2010). Venture capital as a catalyst to commercialization. *Research Policy*, *39*(10), 1348–1360. http://doi.org/10.1016/j.respol.2010.08.006

Sandén, B. A., & Azar, C. (2005). Near-term technology policies for long-term climate targets—economy wide versus technology specific approaches. *Energy Policy*, *33*(12), 1557–1576. http://doi.org/10.1016/j.enpol.2004.01.012

Sanderson, S. W., & Simons, K. L. (2014). Light emitting diodes and the lighting revolution: The emergence of a solid-state lighting industry. *Research Policy*. http://doi.org/10.1016/j.respol.2014.07.011

Sanstad, A. H., & Howarth, R. B. (1994). "Normal" markets, market imperfections and energy efficiency. *Energy Policy*, *22*(10), 811–818. http://doi.org/10.1016/0301-4215(94)90139-2

Santamaría, L., Barge-Gil, A., & Modrego, A. (2010). Public selection and financing of R&D cooperative projects: Credit versus subsidy funding. *Research Policy*, *39*(4), 549–563. http://doi.org/10.1016/j.respol.2010.01.011

Sartorius, C. (2008). Promotion of stationary fuel cells on the basis of subjectively perceived barriers and drivers. *Journal of Cleaner Production*, *16*(1, Supplement 1), S171–S180. http://doi.org/10.1016/j.jclepro.2007.10.013

SBI. (2013). Contracting zur Modernisierung der kommunalen Straßenbeleuchtung. Retrieved from http://licht.cfi21.org

Schilling, M. A., & Esmundo, M. (2009). Technology S-curves in renewable energy alternatives: Analysis and implications for industry and government. *Energy Policy*, *37*(5), 1767–1781. http://doi.org/10.1016/j.enpol.2009.01.004

Schischke, K., Nissen, N. F., Stobbe, L., & Reichl, H. (2008). Energy efficiency meets ecodesign – technology impacts of the European EuP directive. In *IEEE*

International Symposium on Electronics and the Environment, 2008. ISEE 2008 (pp. 1–6). http://doi.org/10.1109/ISEE.2008.4562882

Schleich, J. (2009). Barriers to energy efficiency: A comparison across the German commercial and services sector. *Ecological Economics, 68*(7), 2150–2159. http://doi.org/10.1016/j.ecolecon.2009.02.008

Schmidt, R. C., & Marschinski, R. (2009). A model of technological breakthrough in the renewable energy sector. *Ecological Economics, 69*(2), 435–444. http://doi.org/10.1016/j.ecolecon.2009.08.023

Schönberger, P. (2013). Municipalities as key actors of German renewable energy governance: an analysis of opportunities, obstacles, and multi-level influences. *Wuppertal Papers, 186*. Retrieved from http://epub.wupperinst.org/frontdoor/index/index/docId/4676

Schumpeter, J. A. (1939). *Business cycles* (Vol. 100). Cambridge Univ Press.

Seawright, J., & Gerring, J. (2008). Case Selection Techniques in Case Study Research. *Political Research Quarterly, 61*(2), 294–308. http://doi.org/10.1177/1065912907313077

Selviaridis, K., & Wynstra, F. (2014). Performance-based contracting: a literature review and future research directions. *International Journal of Production Research, 0*(0), 1–36. http://doi.org/10.1080/00207543.2014.978031

Shin, D., Curtis, M., Huisingh, D., & Zwetsloot, G. I. (2008). Development of a sustainability policy model for promoting cleaner production: a knowledge integration approach. *Journal of Cleaner Production, 16*(17), 1823–1837. http://doi.org/10.1016/j.jclepro.2008.06.006

Siggelkow, N. (2007). Persuasion with case studies. *The Academy of Management Journal, 50*(1), 20–24. http://doi.org/10.5465/AMJ.2007.24160882

Sine, W. D., & Lee, B. H. (2009). Tilting at Windmills? The Environmental Movement and the Emergence of the U.S. Wind Energy Sector. *Administrative Science Quarterly, 54*, 123–155. http://doi.org/10.2189/asqu.2009.54.1.123

Smink, M. M., Hekkert, M. P., & Negro, S. O. (2013). Keeping sustainable innovation on a leash? Exploring incumbents' institutional strategies. *Business Strategy and the Environment.* http://doi.org/10.1002/bse.1808

Smith, A., Stirling, A., & Berkhout, F. (2005). The governance of sustainable sociotechnical transitions. *Research Policy, 34*(10), 1491–1510. http://doi.org/10.1016/j.respol.2005.07.005

Smith, A., Voß, J.-P., & Grin, J. (2010). Innovation studies and sustainability transitions: The allure of the multi-level perspective and its challenges. *Research Policy, 39*(4), 435–448. http://doi.org/10.1016/j.respol.2010.01.023

Smith, S., & Swierzbinski, J. (2007). Assessing the performance of the UK Emissions Trading Scheme. *Environmental and Resource Economics, 37*(1). http://doi.org/10.1007/s10640-007-9108-5

Sorrell, S. (2005). *The contribution of energy service contracting to a low carbon economy.* Tyndall Centre for Climate Change Research.

Sorrell, S. (2007). The economics of energy service contracts. *Energy Policy, 35*(1), 507–521. http://doi.org/10.1016/j.enpol.2005.12.009

Sorrell, S., O'Malley, E., Schleich, J., & Scott, S. (2004). *The economics of energy efficiency: barriers to cost-effective investment.* Edward Elgar Cheltenham.

Sovacool, B. K. (2008). Replacing tedium with transformation: Why the US Department of Energy needs to change the way it conducts long-term R&D. *Energy Policy, 36*(3), 923–928. http://doi.org/10.1016/j.enpol.2007.11.018

Sovacool, B. K. (2009a). Rejecting renewables: The socio-technical impediments to renewable electricity in the United States. *Energy Policy, 37*(11), 4500–4513. http://doi.org/10.1016/j.enpol.2009.05.073

Sovacool, B. K. (2009b). The importance of comprehensiveness in renewable electricity and energy-efficiency policy. *Energy Policy, 37*(4), 1529–1541. http://doi.org/10.1016/j.enpol.2008.12.016

Steinbach, A. (2013). Barriers and solutions for expansion of electricity grids— the German experience. *Energy Policy, 63*, 224–229. http://doi.org/10.1016/j.enpol.2013.08.073

Steinberger, J. K., van Niel, J., & Bourg, D. (2009). Profiting from negawatts: Reducing absolute consumption and emissions through a performance-based energy economy. *Energy Policy, 37*(1), 361–370. http://doi.org/10.1016/j.enpol.2008.08.030

Stern, N. (2007). *The economics of climate change: the Stern review.* Cambridge University Press.

Stern, N. (2008). The Economics of Climate Change. *The American Economic Review, 98*(2), 1–37. http://doi.org/10.2307/29729990

Stewart, J., & Hyysalo, S. (2008). Intermediaries, Users and Social Learning in Technological Innovation. *International Journal of Innovation Management, 12*(3), 295–325. http://doi.org/10.1142/S1363919608002035

Strand, J., & Toman, M. (2010). *"Green Stimulus", Economic Recovery, and Long-Term Sustainable Development* (SSRN Scholarly Paper No. ID 1533681). Rochester, NY: Social Science Research Network. Retrieved from http://papers.ssrn.com/abstract=1533681

Szabó, S., & Jäger-Waldau, A. (2008). More competition: Threat or chance for financing renewable electricity? *Energy Policy, 36*(4), 1436–1447. http://doi.org/10.1016/j.enpol.2007.12.020

Szarka, J. (2006). Wind power, policy learning and paradigm change. *Energy Policy, 34*(17), 3041–3048. http://doi.org/10.1016/j.enpol.2005.05.011

Tampakis, S., Tsantopoulos, G., Arabatzis, G., & Rerras, I. (2013). Citizens' views on various forms of energy and their contribution to the environment. *Renewable and Sustainable Energy Reviews, 20*, 473–482. http://doi.org/10.1016/j.rser.2012.12.027

Testa, F., Annunziata, E., Iraldo, F., & Frey, M. (2014). Drawbacks and opportunities of green public procurement: an effective tool for sustainable production. *Journal of Cleaner Production*. Retrieved from http://www.sciencedirect.com/science/article/pii/S0959652614010312

Toffel, M. W. (2002). Contracting for Servicizing. *Harvard Business School Technology and Operations Mgt. Unit Research Paper, 08–063*(2).

Ughetto, E. (2007). The Financing of Innovative Activities By Banking Institutions: Policy Issues and Regulatory Options. *SSRN eLibrary*. Retrieved from http://papers.ssrn.com/sol3/papers.cfm?abstract_id=937207

Ughetto, E. (2010). Assessing the contribution to innovation of private equity investors: A study on European buyouts. *Research Policy, 39*(1), 126–140. http://doi.org/10.1016/j.respol.2009.11.009

Upreti, B. R. (2004). Conflict over biomass energy development in the United Kingdom: some observations and lessons from England and Wales. *Energy Policy, 32*(6), 785–800. http://doi.org/10.1016/S0301-4215(02)00342-7

Uyarra, E., Edler, J., Garcia-Estevez, J., Georghiou, L., & Yeow, J. (2014). Barriers to innovation through public procurement: A supplier perspective. *Technovation*. http://doi.org/10.1016/j.technovation.2014.04.003

van den Bergh, J. C. J. M. (2013). Environmental and climate innovation: Limitations, policies and prices. *Technological Forecasting and Social Change, 80*(1), 11–23. http://doi.org/10.1016/j.techfore.2012.08.004

van Lente, H., Hekkert, M. P., Smits, R., & van Waveren, B. (2003). Roles of Systemic Intermediaries in Transition Processes. *International Journal of Innovation Management, 07*(03), 247–279. http://doi.org/10.1142/S1363919603000817

van Soest, D. P., & Bulte, E. H. (2001). Does the Energy-Efficiency Paradox Exist? Technological Progress and Uncertainty. *Environmental and Resource Economics, 18*(1). http://doi.org/10.1023/A:1011112406964

Verdolini, E., & Galeotti, M. (2011). At home and abroad: An empirical analysis of innovation and diffusion in energy technologies. *Journal of Environmental Economics and Management, 61*(2), 119–134. http://doi.org/10.1016/j.jeem.2010.08.004

Veugelers, R. (2011). *Europe's clean technology investment challenge* (Policy Contributions No. 561). Bruegel. Retrieved from http://ideas.repec.org/p/bre/polcon/561.html

Veugelers, R. (2012). Which policy instruments to induce clean innovating? *Research Policy, 41*(10), 1770–1778. http://doi.org/10.1016/j.respol.2012.06.012

Weber, K. M., & Rohracher, H. (2012). Legitimizing research, technology and innovation policies for transformative change: Combining insights from innovation systems and multi-level perspective in a comprehensive "failures" framework. *Research Policy, 41*(6), 1037–1047. http://doi.org/10.1016/j.respol.2011.10.015

White, W., Lunnan, A., Nybakk, E., & Kulisic, B. (2013). The role of governments in renewable energy: The importance of policy consistency. *Biomass and Bioenergy, 57*, 97–105. http://doi.org/10.1016/j.biombioe.2012.12.035

Wieczorek, A. J., & Hekkert, M. P. (2012). Systemic instruments for systemic innovation problems: A framework for policy makers and innovation scholars. *Science and Public Policy, 39*(1), 74–87. http://doi.org/10.1093/scipol/scr008

Williamson, O. E. (1985). *The Economic Institutions of Capitalism*. New York: Free Press.

Wilson, C., Grubler, A., Gallagher, K. S., & Nemet, G. F. (2012). Marginalization of end-use technologies in energy innovation for climate protection. *Nature Climate Change, 2*(11), 780–788. http://doi.org/10.1038/nclimate1576

Wodak, R. (2004). Critical discourse analysis. In C. Seale, J. F. Gubrium, & D. Silverman (Eds.), *Qualitative Research Practice*. London: Sage.

Wonglimpiyarat, J. (2011). The dynamics of financial innovation system. *The Journal of High Technology Management Research, 22*(1), 36–46. http://doi.org/10.1016/j.hitech.2011.03.003

Wooldridge, J. M., Calhoun, J. W., Jung, B., Greber, S., & Montgomery, R. (2009). *Introductory Econometrics* (4th ed.). Mason, OH: South-Western Cengage Learning.

Wüstenhagen, R., & Menichetti, E. (2012). Strategic choices for renewable energy investment: Conceptual framework and opportunities for further research. *Energy Policy, 40*, 1–10. http://doi.org/10.1016/j.enpol.2011.06.050

Wüstenhagen, R., Wolsink, M., & Bürer, M. J. (2007). Social acceptance of renewable energy innovation: An introduction to the concept. *Energy Policy, 35*(5), 2683–2691. http://doi.org/10.1016/j.enpol.2006.12.001

Yaqub, O., & Nightingale, P. (2012). Vaccine innovation, translational research and the management of knowledge accumulation. *Social Science & Medicine, 75*(12), 2143–2150. http://doi.org/10.1016/j.socscimed.2012.07.023

Yin, R. K. (2009). *Case study research: Design and methods* (Vol. 5). Sage Publications, Inc.

Yusuf, S. (2008). Intermediating knowledge exchange between universities and businesses. *Research Policy, 37*(8), 1167–1174. http://doi.org/10.1016/j.respol.2008.04.011

Zhang, X., Shen, L., & Chan, S. Y. (2012). The diffusion of solar energy use in HK: What are the barriers? *Energy Policy, 41*, 241–249. http://doi.org/10.1016/j.enpol.2011.10.043

Zhang, X., Wu, Z., Feng, Y., & Xu, P. (2014). "Turning green into gold": a framework for energy performance contracting (EPC) in China's real estate industry. *Journal of Cleaner Production.* http://doi.org/10.1016/j.jclepro.2014.09.037

8 Appendix

8.1 Appendix for chapter 3

8.1.1 Supplementary data (i.e. organised quotes)

Table 8-1: Representative quotes for barriers, roles and functions of the intermediaries

Antecedent	Characteristic	Representative quotes
Barriers to low-carbon-innovation	*Technological*	The production technology area is complex. Tasks are complex, equipment is complex and finally we need partners who buy the products, meaning reliable partnerships. (Research project manager, Carbody technologies) (B1)
		I see a deficit in the 'Mittelstand'. How do we make those companies more innovative? (Head of research, Biotechnology) (B8)
	Regulatory / Political	Recently we have been overwhelmed with bureaucracy. Nowadays we need to document why we support mature companies. (Research project manager, Advanced materials) (B2)
		We are limited by rules for granting support, public procurement law etc. We are not allowed to support individual firms but we can take measures that enhance innovation in Germany as a location for businesses. (Research project manager, Smart grids) (B3)
		Infrastructure is a technical and a cost challenge. We have access to comprehensive analyses. (Research project manager, Fuel cells) (B13)
		Regarding the vehicles [the main barriers] are costs. In terms of technology we rather see an optimisation problem which is the case for all new technologies. [Fuel cell cars] are disadvantaged compared to incumbent technologies. The incumbent technologies respond to user demands but ignore other challenges. (Research project manager, Fuel cells) (B15)
		Only political barriers remain for CCS, especially acceptance among the population. Nobody would initiate a demonstration project if it was not safe. (Research project manager, Carbon capture and storage) (B16)
		The cost target is not yet achieved. The question is, what to do in-between [R&D and commercialisation]. Financial incentives to account for the higher upfront investment are desired by industry. [...] However a feed-in tariff as for the renewables would not be adequate. (Research project manager, Fuel cells) (B17)

Antecedent	Characteristic	Representative quotes
	Cooperative	In that case we have to assure confidentiality, especially towards our project partners. We know interesting numbers. […] The goal is to move beyond R&D. (Research project manager, Photovoltaics) (B6)
		As you can imagine, these information [market developments, expertise, evaluations of technologies] are confidential. We only report aggregated data. But the companies also have limitations in terms of what they can share. (Research project manager, Carbon capture and usage technologies) (B9)
		Financiers have built up competences since the 1990s, especially in banks and venture capitalists. However we are not asked for advice. (Head of research, Biotechnology) (B14)
	Financial	We had good projects which we could not fund due to insufficient budgets. (B4)
		The industry is not willing to invest. That is critical. […] We hear this a lot from our project partners (Research project manager, Solar energy technologies) (B5)
		We need to find someone who is willing to invest and bear the risk. We need to find someone who is willing to do that in Germany. I do not think that taxpayers will be excited if we fund a German company which establishes its production facility in Asia, the US or elsewhere. (Research project manager, Battery technologies) (B7)
		The feedback from our demonstration projects, where smart homes are deployed, is that there is only a business case when the pay scale fits the user needs. (Research project manager, Smart grids) (B10)
		R&D is important, but we have realised that apart from supporting technology policy an industrial policy monitoring is necessary. […] What kind of instruments could we use? Loans for market introduction? After the R&D phase, technologies need 5–10 times more investments for commercialisation. Can companies bear this burden? Do they get private finance or not? (Research project manager, Smart grids) (B11)
		The problem is credit-worthiness [for co-financed R&D projects and for commercialisation efforts]. It is a real financing problem. When they apply for a R&D project they need collateral. Sometimes banks give guarantees, but this is rare. There I see the main problem for our companies. (Research project manager, Smart grids) (B12)
		[The company] has left the R&D phase early. At some point in time they did not have enough financial resources. They built a production facility early without actually knowing what they are producing. So the costs where high and the market did not take the products. (Research project manager, Organic LED) (B18)
		Currently [potential buyers of the innovative technology] rather go with the subsidies instead of private energy service contracting. If we do not subsidise the products anymore contracting will be used. (Research project manager, LED) (B19)

Antecedent	Characteristic	Representative quotes
Roles, Competencies and mandates **Policy support**	*Expertise*	It is not the case that we could go into the laboratory ourselves and take over the work there but we know the main research streams and their contents very well. We need to know where the problems are and what the solutions could look like. (Research project manager, Organic LED) (C1)
	Knowledge broker	We see ourselves as brokers. We work for the ministries but somehow also for the industry. That means we attempt to bring together different points of view. That is our strength because we are familiar with different sectors. We are in best sense mediators and we try to advance the opinions of others rather than our own. (Research project manager, Photovoltaic) (C2)
	Technology transfer	We funded these technology centres for ten years. There is a lot of know-how, numerous highly ranked publications and eminently respectable researchers, […] but what we originally intended, the next step, technology transfer into the firms [did not happen] the firms acted very conservatively. (Head of research, Biotechnology program) (C3) That is the mission of most funding programs, it is about technology transfer. On one hand we have industrial R&D which is done autonomously. On the other, we have university R&D, Max-Planck-Society – basic research. The Government tries to close the gap by using project funding and by supporting technology transfer. (Head of research, Biotechnology) (C4)
	Bridge builder	It is our task to make sure that everyone gets a word in edgewise, to ask critical questions and to determine what is important. In general representatives from the corresponding ministry are present. […] It makes sense that they concentrate on content. So in the end it is our duty to collect information, to channelize it and to question what is really needed. (Research project manager, Organic photovoltaic) (C5)
	Manager	I think we do not need to do cost-revenue-control; the firms are doing that already. And we get ambiguous feedback. There are companies saying that it is helpful what you do and others do not feel informed enough. (Research project manager, LED) (C6) We evaluate [projects] according to a matrix, e.g. market perspective 40%, congruency with national fuel cell initiative and sustainability. It is not about proof-of-concept; we want to prepare the market […] in the field of transport and infrastructure. (Research project manager, Fuel cells) (C7)
	Influence on STI policy	It is not that we dictate something […]. We rather make our point and summarise things, which is already an interpretation. But we also get to know political aspects that we are obliged to include in our evaluation. […] We cannot influence it uni-directionally; we rather help them build their opinion. (Research project manager, Geothermal energy) (P1) It is our task to set the stage. We would like to have this technology field and not the other. […] We need visionary systems – what can they look like? […] We sketch what cooperative research projects could look like and what kind of actors should participate and how close to the application phase they should be. (Research project manager, Advanced materials) (P2)

Antecedent	Characteristic	Representative quotes
		Many projects produced meaningful results, but there were no business models. The regulatory framework does not permit to incorporate the results into business models. (Research project manager, Smart grids) (P3)
		[The regulation] is coordinated here, that is a huge advantage. The standardisation took place including discussions where the counter would be installed – at the car or at the motoring pump. That will be a crucial advantage in the future. (Research project manager, Fuel cells) (P4)
		It's about acceptance, not only the tank vs. plate discussion, also whether the introduction of renewable resources and fuels […], could create disadvantages regarding ecology or quality of life. (Research project manager, Biofuels) (P5)
		[The regulatory environment] exhibits positive and negative influence on the innovation behaviour. Sure, companies follow trends if they are clearly visible. Production research respectively the production industry is conservative as the processes are highly complex […]. (Research project manager, Carbody technologies) (P6)
	Interfaces between public authorities	A concrete example was the coordination between ministry X and ministry Y. In the first place the competencies are marked off, ministry X does the research and ministry Y the application. But in the end it is a very fuzzy field. (Research project manager, Lithium-Ion-Battery) (P7)
		I think the most critical point is what happens at the end of projects concerned with basic R&D. We are not responsible anymore because we funded the basic R&D. Another ministry might still not be responsible because the application is not visible. […] You often come across the situation that the innovators do not know about the follow-up process. (Research project manager, Organic photovoltaic) (P8)
	Interfaces between innovation phases	In any case we have to answer the question: What comes afterwards? We work for ministry X and fund the research but we do not work for the department that deals with finance and industrial policy. The transition from funded research to industrial policy is no fast-selling-item. (Research project manager, Fuel cells) (P9)
		We strive for synergies and we try to reduce redundancies. But after all the transition from prototype or demonstration, from the R&D phase towards a mainstream product or service is a responsibility of the innovators. (Research project manager, Carbody technologies) (P10)

Antecedent	Characteristic	Representative quotes
Financial instruments	*Combination of public support and private finance*	There are contacts with private actors. Of course we cannot say this is the partner bank X, Y. That cannot be our task. It is rather the question where the [founders] turn to get own contacts. […] We are willing to establish contacts but we cannot act as handmaidens of a bank. (Research project manager, Biofuels) (F1) […] I would see it as our task that if we cannot fund them that we use our relationships with business angels who know that we have expertise in the field. However many networks do not work, especially not between funding ministries. (Research project manager, Smart-grids) (F2) We have developed [a roadmap] with confidential company data. Especially scenarios for infrastructure and joined these with roll-out scenarios of the automobile industry. That is different to the infrastructure for natural gas which has been conducted by the utilities that do not have an understanding of the main actors, consumers, automobile users and automobile industry. This has been done here in a coordinated way which represents a huge advantage. (Research project manager, Fuel cells) (F3) [Finance] is a support instrument. If the creditworthiness is missing we would have the routine […] to use venture capital or to give venture-capitalists a hint what we regard as innovative so that they get interested […]. We could also show that we expect growth in this sector over the next years or that we see a lead-market. We also show the sphere around, saying that these small companies are in a consortium with large firms. That is important information. One could also invite [private] funds for jury meetings or involve them in accompanying research. […] still this is not included in our mandate. We are limited in terms of budget. (Research project manager, Smart grids) (F4)
	Limitations	You can imagine that information is partially very confidential. That means the results we extract [from the projects] or hand down are more general. There you read medium-term commercialisation etc. But also from the part of the innovators there are limitations on what we can hand down and what not. (Research project manager, Carbon capture and usage technologies) (F5) We work as a contractor for a certain ministry, selling services to them, precisely giving project funding to the innovators. If we had the industry as a contractor at the same time, I would regard this conflict as critical as long as there is no organisational separation between the two processes. (Research project manager, Organic photovoltaic) (F6)
	Complementary services	We supply preliminary seismic data according to model calculations, simulations; we have the opinion that with probability of 95% you will find the geologic conditions necessary for an insurer to insure the last 5%. (Research project manager, Geothermal energy) (F7)

8.1.2 Interview guide

General questions concerning R&D partnership

- What is the main focus of R&D partnership (R&D, commercialisation, diffusion)?
- How did the technology evolve?
- Who are the directly and indirectly participating actors (i.e. firms, research institutes etc.)?
- What are goals of project managing organisation and participants (industrial, academic)?
- How long does the R&D partnership last?
- Please describe the commitment of the participants (monetary and non-monetary).

Management R&D partnership

- Please describe your role regarding tasks, responsibilities, expertise, hierarchy, status and coordination.
- How is the selection of participants and resource allocation happening?
- What kind of criteria did you apply to select the participants (Expertise, commercialisation, competencies, work plan, risks, leverage effect, overall importance, and financial power)?
- Please describe the evaluation of R&D projects especially regarding transparency.
- Please describe your internal and external communication strategy.

Science, technology and innovation (STI) policy

- Which ministries are responsible for this R&D partnership?
- What are their goals?
- What kind of policy measures (e.g. lead markets) do they use?
- Please describe your influence on STI policy.
- Please describe the barriers to commercialisation and diffusion of technology X.
- Which policy instruments could reduce those barriers?

Cooperation

- Please describe the private-private cooperation (Actors, forms, barriers).
- Please describe the public-private cooperation (Actors, forms, barriers).
- Please describe the public-public cooperation (Actors, forms, barriers).

Financing

- Do you think finance is necessary for commercialisation?
- Do you discuss questions related to finance with industrial partners?
- What kind of follow-up finance do they need?
- Have participating actors successfully been (externally) financed?
- Have projects or companies been discontinued due to lack of finance?
- Start-up and growth finance (Business angels, VC, PE; Mezzanine, Banks)
  - Which are questions addressed? Who deals with it?
  - Is there a need for start-up and growth finance? Who are the contacts for the participating actors?
- Please describe your cooperation with financiers.

8.2 Appendix for chapter 4

8.2.1 Interview guide

Modernisation of public street lighting

- How does the process of modernisation unfold in the municipalities?
- Which technologies have been applied in the modernisation process?
- What role did the participating actors (EUCos, MUCos ESCos, manufacturers and financial service providers) play?
- What factors influence their decision making?
- How does the regulatory or political environment influence the modernisation process?

Role of energy service contracts (EPC)

- What are perceived specific success factors and barriers of EPC for LED street lighting in a municipal context?
- How is technological and financial risk treated in these arrangements?
- Can EPC accelerate the diffusion of eco-innovations?

8.2.2 Interview participants

Table 8-2: Sample of lighting and ESCo market actors

Nr	Category	Position	Format	Date	Interviewer
1	Municipal representatives	Lighting engineer	Via telephone	Oct 2013	FP, PvF
2	Municipal representatives	Member of parliament	Via telephone	Oct 2013	FP
3	Municipal representatives	Technical manager	Via telephone	Oct 2013	FP
4	Municipal representatives	Energy efficiency manager	In person	Jan 2014	FP
5	Municipal representatives	Building authority	Via telephone	Oct 2013	FP
6	Municipal representatives	Building authority	Via telephone	Nov 2013	FP
7	Municipal representatives	Lighting manager	In person	Nov 2013	FP
8	Municipal representatives	Building authority	Via telephone	Nov 2013	FP
9	Municipal representatives	Lighting manager	Via telephone	Nov 2013	FP
10	Municipal representatives	Lighting manager	Via telephone	Nov 2013	FP
11	Municipal representatives	Lighting manager	In person	Jan 2014	FP
12	LED Manufacturers	Engineer	Via telephone	Oct 2013	FP, PvF
13	LED Manufacturers	Business developer	In person	Nov 2013	FP
14	LED Manufacturers	Business developer	Via telephone	Dec 2013	FP
15	LED Manufacturers	CEO	In person	Jan 2014	FP
16	LED Manufacturers	Chief marketing officer	In person	Nov 2013	FP
17	Energy service companies ESCos	Business developer lighting	Via telephone	Oct 2013	FP
18	Energy service companies ESCos	Business developer lighting	Via telephone	Oct 2013	FP
19	Energy service companies ESCos	CEO	Via telephone	Nov 2013	PvF
20	Energy service companies ESCos	Business developer lighting	Via telephone	Nov 2013	FP

Nr	Category	Position	Format	Date	Interviewer
21	Energy service companies ESCos	Chief marketing officer	Via telephone	Oct 2013	FP
22	Multi-utility companies MUCos	CEO	Via telephone	Oct 2013	PvF
23	Multi-utility companies MUCos	Lighting manager	Via telephone	Nov 2013	FP
24	Multi-utility companies MUCos	Lighting manager	Via telephone	Dec 2013	FP
25	Multi-utility companies MUCos	CEO	Via telephone	Dec 2013	FP
26	Multi-utility companies MUCos	Lighting manager	In person	Jan 2014	FP
27	Financial service providers	Key account manager for municipal clients	Via telephone	Nov 2013	PvF
28	Financial service providers	Expert on financing energy efficiency projects	Via telephone	Jan 2014	FP, PvF
29	Financial service providers	Key account manager for municipal clients	Via telephone	Jan 2014	FP, PvF
30	Regulatory bodies	Municipal budgetary expert	Via telephone	Jul 2013	FP
31	Regulatory bodies	Municipal budgetary expert	Via telephone	Jul 2013	FP
32	Regulatory bodies	Municipal budgetary expert	Via telephone	Jul 2013	FP
33	Regulatory bodies	Municipal budgetary expert	Via telephone	Oct 2013	FP
34	Facilitators	Energy agency	Via telephone	Oct 2013	FP
35	Facilitators	Energy agency	Via telephone	Oct 2013	FP
36	Facilitators	Public property manager	In person	Dec 2013	PvF
37	Facilitators	Energy consultant	In person	Nov 2013	FP
38	Facilitators	Energy consultant	Via telephone	Dec 2013	FP
39	Facilitators	Municipal agency	Via telephone	Dec 2013	FP
40	Facilitators	Energy agency	Via telephone	Nov 2013	FP

8.3 Appendix for chapter 5

8.3.1 Case selection

Table 8-3: Country selection for the quantitative analysis

Countries included in Multiple RE	Countries included in Solar	Countries included in Wind	Countries included in Biomass
Australia, Austria, Belgium, Canada, Chile, Czech Republic, Denmark, Estonia, Finland, France, Germany, Greece, Hungary, Ireland, Italy, Japan, Korea, Rep., Mexico, Netherlands, New Zealand, Norway, Poland, Portugal, Slovak Republic, Spain, Sweden, Switzerland, Turkey, United Kingdom, United States	Australia, Belgium, Canada, Czech Republic, France, Germany, Greece, Italy, Japan, Korea, Rep., Netherlands, Portugal, Slovak Republic, Spain, Turkey, United Kingdom, United States	Australia, Austria, Belgium, Canada, Chile, Czech Republic, Denmark, Estonia, Finland, France, Germany, Greece, Hungary, Ireland, Italy, Japan, Korea, Rep., Mexico, Netherlands, New Zealand, Norway, Poland, Portugal, Spain, Sweden, Switzerland, Turkey, United Kingdom, United States	Australia, Austria, Belgium, Canada, Chile, Czech Republic, Finland, France, Germany, Ireland, Italy, Japan, Mexico, Netherlands, Norway, Poland, Spain, Sweden, United Kingdom, United States

8.3.2 Summary statistics

Table 8-4: Data – definition, sources and descriptive statistics

Category	Variable	Definition	Source	Obs.	Mean	Std. Dev.	Min	Max
Dependent variables	$BNEF\_Capacity\_ALL_{jk}$	Logarithm of the installed capacity additions of multiple renewable energy sources (aggregated from $BNEF\_Capacity\_Biomass_{jk}$, $BNEF\_Capacity\_Solar_{jk}$, $BNEF\_Capacity\_Wind_{jk}$)	Bloomberg New Energy Finance (BNEF)	360	3.61	2.81	0	9.43
	$BNEF\_Capacity\_Biomass_{jk}$	Logarithm of the installed capacity additions of biomass capacity	idem	240	2.25	2.09	0	6.80
	$BNEF\_Capacity\_Solar_{jk}$	Logarithm of the installed capacity additions of solar capacity	idem	192	1.76	2.27	0	8.56
	$BNEF\_Capacity\_Wind_{jk}$	Logarithm of the installed capacity additions of wind capacity	idem	348	3.34	2.81	0	9.17
Economic Instruments – Fiscal/financial incentives (EI_FI)	$EI\_FI\_FI_{jk}$	Logarithm of ANPM (feed-in tariffs/premiums)	IEA policies and measures database	360	0.48	0.52	0	1.95
	$EI\_FI\_GS_{jk}$	Logarithm of ANPM (grants and subsidies)	idem	360	0.88	0.64	0	2.30
	$EI\_FI\_L_{jk}$	Logarithm of ANPM (loans)	idem	360	0.25	0.38	0	1.61
	$EI\_FI\_TR_{jk}$	Logarithm of ANPM (tax relief)	idem	360	0.40	0.52	0	1.79
	$EI\_FI\_T_{jk}$	Logarithm of ANPM (taxes)	idem	360	0.27	0.37	0	1.10
Economic Instruments – Market-based instruments (EI_MI)	$EI\_MI\_GA_{jk}$	Logarithm of ANPM (GHG emissions allowances)	idem	360	0.10	0.29	0	1.39
	$EI\_MI\_GC_{jk}$	Logarithm of ANPM (green certificates)	idem	360	0.19	0.37	0	1.61

Category	Variable	Definition	Source	Obs.	Mean	Std. Dev.	Min	Max
Economic Instruments – Direct investment (EI_DI)	$EI\_DI\_FSG_{jk}$	Logarithm of ANPM (funds to sub-national governments)	idem	360	0.16	0.32	0	1.39
	$EI\_DI\_II_{jk}$	Logarithm of ANPM (infrastructure investments)	idem	360	0.17	0.36	0	1.60
Policy Support (PS)	PS_IC_{jk}	Logarithm of ANPM (institutional creation)	idem	360	0.47	0.49	0	1.95
	PS_SP_{jk}	Logarithm of ANPM (strategic planning)	idem	360	0.87	0.56	0	2.08
Regulatory Instruments (RI)	RI_CS_{jk}	Logarithm of ANPM (codes and standards)	idem	360	0.51	0.52	0	2.40
	RI_OS_{jk}	Logarithm of ANPM (obligation schemes)	idem	360	0.64	0.51	0	2.08
	RI_MR_{jk}	Logarithm of ANPM (other mandatory requirements)	idem	360	0.46	0.56	0	2.30
Control variables	c_TEC_{jk}	Logarithm of total energy consumption	World bank	360	3.50	1.18	1.23	6.96
	c_CI_{jk}	Logarithm of CO2 intensity – Metric tons of carbon dioxide per thousand year 2005 U.S. dollars GDP	EIA	360	0.34	0.17	0.10	0.99
	c_LIR_{jk}	Logarithm of long-term interest rates	OECD	360	1.52	0.60	0	2.82
	c_SP_{jk}	Logarithm of share prices (index)	OECD	360	4.49	0.49	2.90	5.70
	c_GDP_{jk}	Logarithm of real gross domestic product (in billions U.S. dollars, 2013)	World bank	360	26.72	2.00	0	30.34

Table 8-5: Pairwise correlation coefficients (Multiple RE / aggregated sectors)

Variables	IC	EI_FI_FI	EI_FI_GS	EI_FI_L	EI_FI_TR	EI_FI_T	EI_MI_GA	EI_MI_GC	EI_DI_FSG	EI_DI_II	PS_IC	PS_SP	RI_CS	RI_OS	RI_MR	c_TEC	c_CI	c_LIR	c_SP	c_GDP
IC	1.00																			
EI_FI_FI	0.15	1.00																		
EI_FI_GS	0.26	0.21	1.00																	
EI_FI_L	0.23	0.11	0.17	1.00																
EI_FI_TR	0.33	0.09	0.22	0.34	1.00															
EI_FI_T	0.09	−0.01	0.07	0.28	0.07	1.00														
EI_MI_GA	0.20	−0.26	0.01	0.22	0.08	0.57	1.00													
EI_MI_GC	0.18	−0.17	−0.05	0.19	0.19	0.23	0.27	1.00												
EI_DI_FSG	0.14	−0.14	0.28	0.25	0.23	0.19	0.04	−0.09	1.00											
EI_DI_II	0.11	0.10	0.36	0.23	0.26	0.02	−0.05	0.02	0.06	1.00										
PS_IC	0.18	0.22	0.34	0.32	−0.01	0.19	0.14	−0.14	0.08	0.27	1.00									
PS_SP	0.29	−0.03	0.26	0.34	−0.01	0.34	0.41	0.37	0.26	0.05	0.40	1.00								
RI_CS	0.32	0.10	0.38	0.25	0.44	0.23	0.05	0.22	0.34	0.29	0.27	0.30	1.00							
RI_OS	0.36	0.08	0.23	0.17	0.30	0.01	0.20	−0.00	0.17	0.08	0.46	0.15	0.34	1.00						
RI_MR	0.33	−0.11	0.07	0.41	0.45	0.18	0.08	0.16	0.40	0.26	0.03	0.15	0.40	0.40	1.00					
c_TEC	0.40	−0.12	0.30	0.31	0.38	0.04	0.19	0.06	0.39	0.13	0.19	0.13	0.30	0.45	0.53	1.00				
c_CI	−0.17	−0.04	−0.18	−0.17	−0.10	−0.22	−0.19	−0.13	−0.18	0.08	−0.19	−0.34	−0.16	−0.00	0.01	−0.26	1.00			
c_LIR	−0.01	0.08	0.16	0.16	0.15	0.25	0.10	0.13	0.09	0.17	0.16	0.18	0.22	0.12	0.11	0.05	−0.28	1.00		
c_SP	0.31	0.27	0.19	0.21	0.20	0.19	0.05	0.22	0.08	0.01	0.26	0.37	0.24	0.22	0.10	0.07	−0.33	0.24	1.00	
c_GDP	0.37	0.02	0.23	0.21	0.32	0.09	0.16	0.16	0.17	0.10	0.10	0.10	0.31	0.33	0.39	0.70	−0.31	0.07	0.16	1.00

8.3.3 Robustness checks

Table 8-6: Random effects estimator (REE) regression results

Independent variables (ANPM)$_{(t-1)}$	Random effects							
	CSE		CSE		CSE		CSE	
	Multiple RE		Wind		Solar		Biomass	
	(IX)		(X)		(XI)		(XII)	
	Coeff.	S.E.	Coeff.	S.E.	Coeff.	S.E.	Coeff.	S.E.
EI_FI_FI$_{jk}$	0.65*	0.38	0.61*	0.36	1.18***	0.31	0.09	0.26
EI_FI_GS$_{jk}$	0.98***	0.36	0.83**	0.36	0.61**	0.25	0.61***	0.25
EI_FI_L$_{jk}$	−0.38	0.55	0.66	0.53	−0.91**	0.43	−0.15	0.39
EI_FI_TR$_{jk}$	0.36	0.43	−0.02	0.42	−0.01	0.34	0.10	0.27
EI_FI_T$_{jk}$	−1.43**	0.59	−0.50	0.58	1.00*	0.58	−0.20	0.39
EI_MI_GA$_{jk}$	2.06***	0.80	1.46*	0.83	−1.17	0.80	0.97**	0.53
EI_MI_GC$_{jk}$	0.15	0.56	0.24	0.52	−2.27***	0.48	0.06	0.37
EI_DI_FSG$_{jk}$	−1.12	0.71	−1.11	0.70	−1.43***	0.47	1.51***	0.52
EI_DI_II$_{jk}$	0.12	0.52	0.52	0.46	0.21	0.56	−1.20***	0.46
PS_IC$_{jk}$	−0.55	0.48	0.02	0.45	−1.51***	0.43	0.83**	0.36
PS_SP$_{jk}$	1.20***	0.42	0.66	0.43	2.35***	0.34	−0.44	0.30
RI_CS$_{jk}$	0.82*	0.48	1.03**	0.47	0.54	0.36	−0.41	0.33
RI_OS$_{jk}$	0.35	0.48	0.26	0.47	0.16	0.41	−0.01	0.34
RI_MR$_{jk}$	0.73	0.46	0.04	0.43	0.77	0.47	0.28	0.31
Control variables								
c_TEC$_{jk}$	0.62**	0.28	0.83	0.30	0.95	0.72	−1.97***	0.52
c_CI$_{jk}$	−0.66	1.60	−0.38	1.80	−6.01**	2.58	4.37***	1.60
c_LIR$_{jk}$	−0.71**	0.31	−0.75**	0.31	−0.54	0.67	−0.57*	0.30
c_SP$_{jk}$	1.84***	0.36	1.92***	0.37	−0.24	0.46	0.61*	0.37
c_GDP$_{jk}$	−0.04	0.08	−0.05	0.08	−0.82	0.70	2.16***	0.50
_cons	−7.54***	2.79	−8.64***	2.80	21.59	17.02	−53.24***	11.51
Observations	330		319		176		220	
R^2	0.36		0.36		0.49		0.38	
Wald	205.74***		198.02***		152.19***		123.86***	

Notes: The Wald test has a Chi^2 distribution and tests the null hypothesis of non-significance of all coefficients of independent variables. Conventional standard errors (CSE) are reported. ***, **, *, denote significance at 1, 5 and 10% significance levels, Estimations include both country and time dummies. (Marques and Fuinhas, 2012a) xtreg command was used.

Finanzmärkte und Klimawandel

Herausgegeben von Dirk Schiereck und Paschen von Flotow

Band 1 Christian Babl / Paschen von Flotow / Dirk Schiereck (Hrsg.): Projektrisiken und Finanzie-
rungsstrukturen bei Investitionen in erneuerbare Energien. 2011.

Band 2 Christoph Ettenhuber: Financing Corporate Growth in the Renewable Energy Industry. 2013.

Band 3 Anette von Ahsen / Robert Fraunhoffer / Dirk Schiereck (Hrsg.): Wellenbrecher auf dem
Weg zur Energiewende? Zur Attraktivität von Energiespeicherung, nachhaltiger Erzeugung
und Verbrauchersteuerung. 2013.

Band 4 Christian Friebe: Diffusion of Renewable Energy Technologies. Private Sector Perspectives on Emerging Markets. 2014.

Band 5 Paschen von Flotow / Dirk Schiereck / Julian Trillig (Hrsg.): Energietransformation, dezent-
rale Erzeugungsprobleme und Finanzierung der Solarindustrie. 2014.

Band 6 Christian Babl: E-Mobility and Related Clean Technologies from an Empirical Corporate
Finance Perspective. State of Economic Research, Sourcing Risks, and Capital Market
Perception. 2015.

Band 7 Friedemann Polzin: Addressing Barriers to Low-Carbon Innovation. Essays on Structures
and Policies to Mobilise Private Finance. 2015.

www.peterlang.com